Studies in Logic
Mathematical Logic and Foundations
Volume 22

The Axiom of Choice

Volume 11
Foundations of the Formal Sciences V: Infinite Games
Stefan Bold, Benedikt Löwe, Thoralf Räsch and Johan van Benthem, editors

Volume 12
Second-Order Quantifier Elimination: Foundations, Computational Aspects and Applications
Dov M. Gabbay, Renate A. Schmidt and Andrzej Szałas

Volume 13
Knowledge in Flux. Modeling the Dynamics of Epistemic States
Peter Gärdenfors. With a foreword by David Makinson

Volume 14
New Approaches to Classes and Concepts
Klaus Robering, editor

Volume 15
Logic, Navya-Nyāya and Applications. Homage to Bimal Krishna Matilal
Mihir K. Chakraborti, Benedikt Löwe, Madhabendra Nath Mitra and Sundar Sarukka

Volume 16
Foundations of the Formal Sciences VI. Probabilistic Reasoning and Reasoning with Probabilities.
Benedikt Löwe, Eric Pacuit and Jan-Willem Romejin, eds.

Volume 17
Reasoning in Simple Type Theory. Festschrift in Honour of Peter B. Andrews on His 70th Birthday.
Christoph Benzmüller, Chad E. Brown and Jörg Siekmann, eds.

Volume 18
Classification Theory for Abstract Elementary Classes
Saharon Shelah

Volume 19
The Foundations of Mathematics
Kenneth Kunen

Volume 20
Classification Theory for Abstract Elementary Classes, Volume 2
Saharon Shelah

Volume 21
The Many Sides of Logic
Walter Carnielli, Marcelo E. Coniglio, Itala M. Loffredo D'Ottaviano, eds.

Volume 22
The Axiom of Choice
John L. Bell

Studies in Logic Series Editor

The Axiom of Choice

John L. Bell

ISBN 978-1-904987-54-3

College Publications
Scientific Director: Dov Gabbay
Managing Director: Jane Spurr
Department of Computer Science
King's College London, Strand, London WC2R 2LS, UK

http://www.collegepublications.co.uk

Original cover design by orchid creative www.orchidcreative.co.uk
Printed by Lightning Source, Milton Keynes, UK

Dedicated to the Memory of

My Dear Wife Mimi

1947-2009

ERNST ZERMELO
1871 – 1953

FELIX HAUSDORFF
1868 – 1942

MAX ZORN
1906 – 1993

Preface

The Axiom of Choice has fascinated me from my student days. As a young aspiring mathematician, I was struck by the fact that through the areas of mathematics to which I was most attracted—abstract algebra, general topology, functional analysis —there ran a common thread: the Axiom of Choice and the "working mathematician's" surrogate for it known as Zorn's Lemma. Teasing out that thread is the purpose of the present book.

The book provides an overview of the development of the Axiom of Choice since its introduction by Zermelo at the beginning of the last century. In it the Axiom of Choice is surveyed from three perspectives. The first, or *mathematical* perspective, is that of the abovementioned "working mathematician". This perspective brings into view the manifold applications of the Axiom of Choice—usually in the guise of Zorn`s Lemma—in a great variety of areas of mathematics. The second, *foundational,* perspective, is that of the logician or constructive mathematician concerned with the foundational status of the Axiom of Choice. The third, *topos-theoretical,* perspective is that taken by the mathematician or logician investigating the role of the Axiom of Choice in topos theory.

In this book certain topics—for instance mathematical applications of the Axiom, and its relationship with logic—are discussed in considerable detail. Others—notably the consistency and independence of the Axiom of the usual systems of set theory—are given no more than summary treatment, the justification being that these topics have been given full expositions elsewhere. The book contains seven chapters and two appendices. Chapter I describes the origins of the Axiom of Choice and its status within set theory. Chapter II introduces maximal principles, in particular, Zorn's Lemma, and discusses their development and their relationship to the Axiom of Choice. Chapter III provides a detailed account of the many applications of the Axiom of Choice and Zorn's Lemma within classical mathematics. Chapter IV offers a compressed account of how the Axiom of Choice is proved consistent with, and independent of, Zermelo-Fraenkel set theory. Chapter V describes the relationship between the Axiom of Choice and logic, chiefly focussing attention on intuitionistic logic. Chapter VI begins with a discussion of the Axiom of Choice in a categorical setting, and continues with what amounts to a crash course in topos theory and the associated local set theory, leading up to an account of the role played by the Axiom of Choice therein. Chapter VII is devoted to a discussion of the role of the Axiom of Choice in Constructive Type Theory. The book concludes with two Appendices, the first outlining intuitionistic logic and the second charting the basic concepts of category theory.

*

I wish to thank my old friends and fellow-logicians Peter Aczel and John Mayberry for their helpful comments on the first draft of the book.

Contents

Introduction

The principle of set theory known as the *Axiom of Choice* (**AC**)[1] has been hailed as "probably the most interesting and, in spite of its late appearance, the most discussed axiom of mathematics, second only to Euclid's Axiom of Parallels which was introduced more than two thousand years ago."[2] From this description one might expect **AC** to prove to be as startling an assertion as, say, the Principle of the Constancy of the Velocity of Light or the Quantum Uncertainty Principle. But, unlike the Axiom of Parallels, in its actual formulation **AC** seems humdrum, almost self-evident. As stated by Zermelo in 1904 it amounts to no more than the claim that, given any family $\mathscr{S}$ of nonempty sets, it is possible to select a single element from each member of $\mathscr{S}$. More formally, let us term a *choice function* on $\mathscr{S}$ to be a function f with domain $\mathscr{S}$ such that, for each nonempty set X in $\mathscr{S}$, $f(X)$ is an element of X. Then Zermelo's 1904 formulation of **AC** is the assertion that, for any family $\mathscr{S}$ of nonempty sets, there is at least one choice function on $\mathscr{S}$. If $\mathscr{S}$ is *finite*, the existence of a choice function on $\mathscr{S}$ is a straightforward consequence of the basic principles of set formation and the rules of (classical) logic. When $\mathscr{S}$ is *infinite*, however, these principles no longer suffice and so the existence of a choice function on $\mathscr{S}$ must be the subject of postulation.

---

1 Throughout this book we shall use **AC** as an abbreviation for the Axiom of Choice.

2 Fraenkel, Bar-Hillel and Levy [1973]. It is not quite correct, however, to refer to Euclid's parallel principle as an "Axiom". In the Greek sense the parallel principle is a *postulate* rather than an axiom, and, and as we shall see, the Axiom of Choice may be construed as both an axiom and a postulate.

Zermelo's purpose in introducing **AC** was to establish a central principle of Cantor's set theory, namely, that every set admits a well-ordering and so can also be assigned a cardinal number. The boldness of Zermelo's proposal excited considerable comment from the mathematicians of the day: while **AC** seems to assert the possibility of making indefinitely many arbitrary "choices" — or at least of crystallizing such an imagined procedure into a genuine function—it provides no indication whatsoever of how these "choices" are to be made, or how the resulting function is to be defined. Thus the scepticism of the French mathematician Émile Borel concerning such a possibility was sufficient to move him to declare that "any argument where one supposes an *arbitrary choice* a non-denumerably infinite number of times is outside the domain of mathematics."

In response to these and other criticisms, in 1908 Zermelo offered a formulation of **AC** along with a derivation of the well-ordering principle therefrom, couched in somewhat different terms from that given in his earlier paper. At the same time he made explicit the set-theoretic assumptions underlying his proof, codifying these in the form of postulates which constituted the first axiom system for set theory. These moves did not, however, succeed in silencing his sterner critics.

The tenability of **AC** was later questioned on the grounds that it had "paradoxical" consequences. In 1914 Hausdorff derived from it the startling result that the surface S of a sphere can be decomposed into disjoint sets $S = A \cup B \cup C \cup Q$ in such a way that A, B, C and $B \cup C$ are mutually congruent and Q is countable. In succinct terms, **AC** implies that two-thirds of the surface of a sphere is congruent to one-third of it. In 1924 Banach and Tarski extended Hausdorff's work to three dimensions by showing that any solid sphere can be decomposed into finitely many (later shown by Raphael Robinson to be reducible to 5!) subsets which can themselves be reassembled to form *two* solid spheres, *each of the same size as the original.* They also established that **AC** yields another version of the "paradox", namely, given any pair of solid spheres, either one of them can be decomposed into finitely many subsets which can be reassembled to form a solid sphere of the same size as the other. To put it graphically, **AC** implies that *a sphere the size of the sun can be decomposed and the pieces reassembled so as to form a sphere the size of a pea.*

Despite the "paradoxical" consequences of **AC**, in 1938 Gödel succeeded in establishing its relative consistency with respect to the usual systems of set theory, and this, coupled with its indispensability in the proofs of many significant mathematical theorems, eventually led, if only on pragmatic grounds, to its acceptance by the majority of mathematicians.

Judging by the vast number of its mathematical consequences, **AC** is unquestionably the most fertile principle of set theory. Remarkably, many of these consequences turn out to be *formally equivalent* to it: more than 200 of these equivalents have been recorded. Among the most significant of these equivalents are:

Zermelo's well-ordering theorem: every set can be well-ordered;

Trichotomy Principle: of any pair of cardinal numbers, one is less than the other, or they are equal;

The Kuratowski-Zorn Lemma: any nonempty partially ordered set in which each totally ordered subset has an upper bound posseses a maximal element;

Tychonov's theorem: the product of any family of compact topological spaces is compact;

The model existence theorem for first-order logic: every infinite consistent set Σ of first-order sentences has a model of cardinality no greater than that of Σ;

The Hamel basis theorem: every vector space has a basis.

While the (relative) consistency of **AC** was not established until almost four decades after its formulation, the first steps in confirming its *formal independence* of the basic axioms of set theory were taken by A. Fraenkel as early as 1922. He showed that **AC** is independent of a certain system of set theory allowing the presence of *atoms,* that is, objects possessing no members, yet not identical with the empty set. Remarkable as this advance was, however, it neither answered the question of whether **AC** is independent of the full set-theoretic system of Zermelo-Fraenkel, nor did it demonstrate the independence of the most important consequence of Zermelo's original invocation of **AC**, namely, the existence of a well-ordering of the set of *real numbers*. The issue was finally resolved in 1964 when P. J. Cohen devised his method of *forcing*. Cohen in fact established the independence of a surprisingly weak form of **AC**, namely that asserting the existence of a choice function on a countable family of pairs. Subsequent work by R. M. Solovay and others has established the independence of certain important consequences of **AC**, notably, the Hahn-Banach theorem and the existence of non-Lebesgue measurable sets of real numbers.

Recent work has shown that **AC** plays an even more central role in mathematics and its foundations than was traditionally acknowledged. In 1975, R. Diaconescu, building on ideas of F. W. Lawvere, proved within a category-theoretic setting a result which essentially showed that the classical logical Law of Excluded Middle—the assertion that each proposition is either true or false—can be derived within intuitionistic set theory (in which that law is not assumed) augmented by **AC.** Put succinctly, **AC** *implies the Law of Excluded Middle*. It was later shown that this pivotal law of classical logic can be derived just from the classically trivial version of **AC** that any family of sets with at most two members has a choice function. It is quite remarkable that a combinatorial principle can yield a law of logic.

A new twist in the story of **AC** has recently occurred in connection with the development of systems of constructive[1] mathematics, in particular Martin-Löf's Constructive Type Theory. This can be most easily described by considering the following equivalent form of **AC**:

(*) *for any relation R between sets A, B,*

$$\forall x \in A \exists y \in B\ R(x,y) \Rightarrow \exists f\colon A \to B\ \forall x \in A\ R(x, fx).$$

Now under the constructive interpretation of quantifiers implicit in constructive mathematics, and later given explicit form in Constructive Type Theory, the assertability of an alternation of quantifiers $\forall x \exists y R(x,y)$ means *precisely* that one is given a function f for which $R(x,fx)$ holds for all x. It follows that **AC** in the form (*) is actually *derivable* in such constructive settings. On the other hand this is decidedly not the case for the Law of Excluded Middle. At first sight this seems to clash with the derivability of the Law of Excluded Middle from **AC** in intuitionistic set theory. But it turns out that for the derivation of the Law of Excluded Middle from **AC** to go through it is necessary that sets or functions be *extensional*—that is, are wholly determined by their elements or values. This condition is built into the usual set

[1] In this book the term "constructive" will normally be used in the sense of "compatible with the rules of intuitionistic logic". There is a stricter construal of the term, associated, for example, with Constructive Type Theory, which, in addition to the adherence to intuitionistic logic, also demands the avoidance of impredicative definition. On the rare occasions we need to draw attention to this narrower rendering, we shall use the term "strictly constructive".

theories but is incompatible with Constructive Type Theory. Another condition, formally independent of extensionality, which ensures that the derivation of the Law of Excluded Middle from **AC** goes through is that any equivalence relation determines a quotient set. This is the basic set-theoretic procedure of moving from an equivalence relation to the associated set of "equivalence classes", which amounts to the reduction of equivalence to literal identity. The Law of Excluded Middle can also be shown to follow from a suitably extensionalized version of **AC** itself[1]. The arguments establishing these intriguing results reveal a novel, subtle interplay between **AC** and some of the most fundamental concepts of mathematics and logic. These arguments were originally formulated within Constructive Type Theory, but as is shown in Chapter V of this book, analogous results can be established within a more familiar (to most mathematicians and analytic philosophers at least) *set-theoretic* framework. The core principles of this framework form a theory - *weak set theory* - which lacks the Axiom of Extensionality and supports only minimal set-theoretic constructions. In particular, just as for Constructive Type Theory, within weak set theory the derivation of the Law of Excluded Middle from **AC** cannot be carried out. But, again as with Constructive Type Theory, augmenting weak set theory with extensionality principles or quotient sets enables the derivation to go through.

It seems fair to say that the reputation of **AC** as "probably the most interesting axiom of mathematics" remains undimmed.

[1] In fact, a form essentially amounting to that given by Zermelo in 1908. See Chapter VII.

I
The Axiom of Choice: Its Origins and Status within Set Theory

THE ORIGINS OF AC

In 1904 Ernst Zermelo formulated **AC**[1] in terms of what he called (in English translation) *coverings*. He starts with an arbitrary set *M* (German *Menge*: "set") and uses the symbol *M′* to denote an arbitrary nonempty subset of *M*; the collection of all these latter he denotes by M. He continues:

> *Imagine that with every subset M′ there is associated an arbitrary element* m_1'*, that occurs in M′ itself; let* m_1' *be called the "distinguished" element of M′. This yields a "covering" γ of the set* M *by certain elements of the set M. The number of these coverings is equal to the product* [of the cardinalities of all the subsets *M′*] *and is certainly different from 0.*

The last sentence of this quotation—which asserts, in effect, that coverings always exist for the collection of nonempty subsets of any (nonempty) set—is, in essence, Zermelo's first formulation of **AC**, although he does not give the principle an explicit name at this point. In inviting one to "imagine" a covering it might seem that Zermelo was engaged in mere speculation. But he next asserts that "the number of these coverings is certainly different from 0" on what, given the usual understanding of the term "number", seem to be objective combinatorial grounds—presumably in much the same way that, given a concrete set of, say, 3 elements, one sees immediately that the set of coverings is surely different from 0, (and, with a bit of effort, in fact consists of

[1] Zermelo [1904].

precisely 18 elements). Thus it seems likely that, right from the start, Zermelo regarded **AC** as an essentially combinatorial, "objective" principle governing the concept of set as conceived by Cantor, who, in extending the ancient Greek conception of number, had seen sets as pluralities of individuals, "numbers" of distinct things.

Zermelo's first formulation of **AC** is now usually stated in terms of *choice functions*: here a choice function on a collection $\mathscr{S}$ of nonempty sets is a map f with domain $\mathscr{S}$ such that $f(X) \in X$ for every $X \in \mathscr{S}$. Zermelo's first formulation of **AC** then reads:

AC1 *Any collection of nonempty sets has a choice function.*

In introducing **AC1** Zermelo's purpose was to establish a central principle of Cantor's set theory, namely, that every set admits a well-ordering and so can also be assigned a cardinal number. Zermelo's introduction of **AC**, as well as the use to which he put it, provoked considerable criticism from the mathematicians of the day. The chief objection raised was to what some saw as its highly non-constructive, even idealist, character: while **AC** asserts the possibility of making a number of—perhaps even uncountably many—arbitrary "choices", it gives no indication whatsoever of how these latter are actually to be effected, of how, otherwise put, choice functions are to be *defined.* For this reason Bertrand Russell regarded the principle as dubious at best. The French Empiricists Baire, Borel and Lebesgue, for whom a mathematical object could be asserted to exist only if it can be uniquely defined, went further in explicitly repudiating the principle in the uncountable case[1].

[1] Still, a number of mathematicians came to **regard** **AC** as being true *a priori.* These all broadly shared the view that for a mathematical entity to exist it was not necessary that it be uniquely definable. In [1904] Zermelo himself calls **AC** a "logical principle" which "cannot ... be reduced to a still simpler one" but which, nevertheless, "is applied

In response to these and other criticisms, in 1908 Zermelo offered[1] a formulation of **AC** (and a derivation of the well-ordering principle therefrom) couched in somewhat different terms from that given in his earlier paper. He uses the term "postulate of choice" or "general principle of choice" to refer to the principle introduced there, which he now formulates as follows:

> *a simultaneous choice of distinguished elements is in principle always possible for an arbitrary set of sets, or, to be more precise, ... the same consequences hold as if such a choice were possible.*

He goes on to admit that, in this formulation, the principle still appears to be "somewhat tainted with subjectivity". To remedy this he proposes to replace it with (or "reduce it to") the following

> *AXIOM. A set S that can be decomposed into a set of disjoint parts A, B, C, ... , each containing at least one element, possesses at least one subset S_1 having exactly one element with each of the parts A, B, C, ... , considered.*

Of this Axiom he observes that its "purely objective character is immediately evident." Perhaps Zermelo regarded the move from his 1904 version of **AC**, with its "taint of subjectivity" to its "objective" 1908 formulation as the transformation of a mere postulate into a true axiom.

Let us call a *transversal* for a family of sets $\mathscr{S}$ any subset $T \subseteq \bigcup\mathscr{S}$ for which each intersection $T \cap X$ for $X \in \mathscr{S}$ has exactly one element. Zermelo's 1908 version of the axiom then amounts to

without hesitation everywhere in mathematical deductions." F. P. Ramsey asserts that "the Multiplicative Axiom seems to me the most evident tautology" (Ramsey 1926) . Hilbert employed **AC** in his defence of classical mathematical reasoning against the attacks of the intuitionists: indeed his ε-operators are essentially just choice functions. For him, "the essential idea on which the axiom of choice is based constitutes a general logical principle which, even for the first elements of mathematical inference, is indispensable" (Hilbert 1926).

[1] Zermelo [1908].

the assertion that any family of mutually disjoint nonempty sets has a transversal.

In claiming that his new axiom possesses a "purely objective character", Zermelo seems to have intended to emphasize the fact that in this form the principle makes no appeal to the idea of making "choices", whose presence in its original formulation had excited so much criticism. It may also be that Zermelo had something like the following "combinatorial" justification of the principle in mind. Given a family $\mathscr{S}$ of mutually disjoint nonempty sets, call a subset $S \subseteq \bigcup\mathscr{S}$ a *cross-section* of $\mathscr{S}$ if $S \cap X \neq \varnothing$ for all $X \in \mathscr{S}$. Clearly cross-sections of $\mathscr{S}$ exist; $\bigcup\mathscr{S}$ itself is an example. Now one can imagine taking a cross-section of S of $\mathscr{S}$ and "thinning out" each intersection $S \cap X$ for $X \in \mathscr{S}$ until it contains just a single element. The result[1] is a transversal for $\mathscr{S}$.

Let us accordingly call Zermelo's 1908 version of **AC** the *Combinatorial* Axiom of Choice:

CAC[2] *Any collection of mutually disjoint nonempty sets has a transversal.*

It is to be noted that **AC1** and **CAC** for *finite* collections of sets are both provable (by induction) in the usual set theories.

AC1 can be reformulated in terms of *indexed sets.* Given an indexed family of sets $\mathfrak{A} = \{A_i : i \in I\}$, each A_i may be conceived of as the "value" of the indexed set **A** at *stage i*. A *choice function* on $\mathfrak{A}$ is a map $f: I \to \bigcup_{i \in I} A_i$ such that $f(i) \in A_i$ for all $i \in I$. A choice

[1] This argument, suitably refined, yields a rigorous derivation of **AC** in this formulation from Zorn's lemma (see Chapter II)

[2] It is this formulation of **AC** that Russell and others refer to as the *multiplicative axiom,* since it is easily seen to be equivalent to the assertion that the product of arbitrary nonzero cardinal numbers is nonzero.

function on $\mathcal{A}$ thus "chooses" an element of the indexed set $\mathcal{A}$ at each stage; a choice function on $\mathcal{A}$ is thus, as it were, an *indexed element* of $\mathcal{A}$. **AC1** is then equivalent to the assertion:

AC2 *Any indexed family of nonempty sets has a choice function.*

Metaphorically speaking, **AC2** amounts to the assertion that an indexed set with an element at each stage has an indexed element.

The set of choice functions on $\mathcal{A}$ is identical with the *product* $\prod_{i\in I} A_i$ of the indexed family $\{A_i: i \in I\}$. Thus **AC2** may also be written in the form *if, for each* $i \in I$, $A_i \neq \varnothing$, *then* $\prod_{i\in I} A_i \neq \varnothing$.

AC1 can also be reformulated in terms of *relations*, viz.

AC3 *for any relation R between sets A, B,*

$$\forall x{\in}A\,\exists y{\in}B\; R(x,y) \Rightarrow \exists f\colon A \to B\; \forall x{\in}A\; R(x, fx).$$

In fact it is easily shown that **AC3** is equivalent to its special case in which A coincides with B, that is,

AC3* *for any binary relation R on a set A,*

$$\forall x \exists y\; R(x,y) \Rightarrow \exists f\colon A \to A\; \forall x\; R(x, fx).$$

Three other equivalent formulations of **AC1** are:

AC4. *Every surjective function has a right inverse.*

AC4*. *For any set* $X \neq \varnothing$ *and any function* $f\colon X \to Y$, there is *a function* $g\colon Y \to X$ *such that* $fgf = f$.

AC5. *Unique representatives can be picked from the equivalence classes of any given equivalence relation.*[1]

THE INDEPENDENCE AND CONSISTENCY OF AC WITHIN SET THEORY

Although the debate concerning **AC** rumbled on for some time, it soon became apparent that the proofs of a number of significant mathematical theorems made essential use of it, so leading many

[1] In this connection we recall Bishop's [1967] observation that *the axiom of choice is used [in classical mathematics] to extract elements from equivalence classes where they should never have been put in the first place.*

mathematicians to adopt it as an indispensable tool of their trade. But while the usefulness of **AC** quickly become clear, doubts concerning its soundness remained. These doubts were compounded by the discovery that **AC** had a number of highly counterintuitive geometrical consequences, the most spectacular of which was Banach and Tarski's[1] *paradoxical decompositions of the sphere*. They showed that, under the assumption of **AC,** any solid sphere can be split into finitely many pieces which can be reassembled to form two solid spheres of the same size; and any solid sphere can be split into finitely many pieces in such a way as to enable them to be reassembled to form a solid sphere of arbitrary size.

There was also the question of **AC**'s *independence* of the system of set-theoretic axioms that Zermelo had put forward in 1908[2]. It was in connection with this problem that the first major advance was made in 1922 when Fraenkel proved the independence of **AC** from a system of set theory containing "atoms". Here by an *atom* is meant a pure individual, that is, an entity having no members and yet distinct from the empty set (so *a fortiori* an atom cannot be a set). In a system of set theory with atoms it is assumed that one is given an infinite set A of atoms. That being the case, one can build a universe $V(A)$ of sets over A by starting with A, adding all the subsets of A, adjoining all the subsets of the result, etc., and iterating transfinitely. $V(A)$ is then a model of set theory with atoms. The kernel of Fraenkel's method for proving the independence of **AC** is the observation that, since atoms cannot be set-theoretically distinguished, any permutation of the set A of atoms induces a structure-preserving permutation—an *automorphism*—of the universe $V(A)$ of sets built

[1] Banach and Tarski [1924].

[2] Zermelo [1908a]

from A. This idea may be used to construct another model $Sym(V)$ of set theory—a *permutation* or *symmetric model*—in which a mutually disjoint set of pairs of elements of A has no choice function[1].

Now suppose that we are given a group G of automorphisms of A. Let us say that an automorphism π of A *fixes* an element x of $V(A)$ if $\pi(x) = x$. Clearly, if $\pi \in G$ fixes every element of A, it also fixes every element of $V(A)$. Now it may be the case that, for certain elements $x \in V(A)$, the fixing of the elements of a *subset* of A by any $\pi \in G$ suffices to fix x. We are therefore led to define a *support* for x to be a subset X of A such that, whenever $\pi \in G$ fixes each member of X, it also fixes x. Members of $V(A)$ possessing a *finite* support are called *symmetric.*

We next define the universe $Sym(V)$ to consist of the *hereditarily symmetric* members of $V(A)$, that is, those $x \in V(A)$ such that x, the elements of x, the elements of elements of x, etc., are all symmetric. $Sym(V)$ is also a model of set theory with set of atoms A, and π induces an automorphism of $Sym(V)$.

Now suppose A to be partitioned into a (necessarily infinite) mutually disjoint set P of pairs. Take G to be the group of permutations of A which fix all the pairs in P. Then $P \in Sym(V)$; it can now be shown that $Sym(V)$ contains no choice function on P. For suppose f were a choice function on P and $f \in Sym(V)$. Then f has a finite support which may be taken to be of the form $\{a_1, ..., a_n, b_1, ..., b_n\}$ with each pair $\{a_i, b_i\} \in P$. Since P is infinite, we may select a pair $\{c, d\} = U$ from P different from all the $\{a_i, b_i\}$. Now we define $\pi \in G$ so that π fixes each a_i and b_i and interchanges c and d. Then π also fixes f. Since f was assumed to be a choice function on P, and $U \in P$, we must have $f(U) \in U$, that is,

[1] For a full account of permutation models, see Jech [1973].

$f(U) = c$ or $f(U) = d$. Since π interchanges c and d, it follows that $\pi(f(U)) \neq f(U)$. But since π is an automorphism, it also preserves function application, so that $\pi(f(U)) = \pi f(\pi(U))$. But $\pi(U) = U$ and $\pi f = f$, whence $\pi(f(U)) = f(U)$. We have duly arrived at a contradiction, showing that the universe $Sym(V)$ contains no choice function on P.

The point here is that for a symmetric function f defined on P there is a finite list L of pairs from P the fixing of all of whose elements suffices to fix f, and hence also all the values of f. Now, for any pair U in P but not in L, a permutation π can always be found which fixes all the elements of the pairs in L, but does not fix the members of U. Since π must fix the value of f at U, that value cannot lie in U. Therefore f cannot "choose" an element of U, so *a fortiori* f cannot be a choice function on P.

This argument shows that collections of *sets of atoms* need not necessarily have choice functions, but it fails to establish the same fact for the "usual" sets of mathematics, for example the set of real numbers. That had to wait until 1963 when Paul Cohen showed that it is consistent with the standard axioms of set theory (which preclude the existence of atoms) to assume that a countable collection of pairs of sets of real numbers can fail to have a choice function[1]. The core of Cohen's method of proof[2]—the celebrated method of *forcing*—was vastly more general than any previous technique; nevertheless his independence proof also made essential use of permutation and symmetry in essentially the form in which Fraenkel had originally employed them. Cohen's method was later applied to establish the independence

[1] Notice that any collection of pairs of real numbers has a choice function, since from each pair one may "choose" the lesser of its two elements.

[2] For a full account of Cohen's method of proof, see Bell [2005] or Jech [1973]. A compressed account is offered in Chapter IV of the present book.

of **AC** from "weaker" versions of it such as the Axiom of Dependent Choices and the Boolean Prime Ideal Theorem, as well as the independence of these weaker versions from the standard axioms of set theory.

Fraenkel's and Cohen's methods of demonstrating the independence of **AC** both rest on the idea of *enlarging* the universe of sets *V* to a universe *V′* in which a new permutation has been "adjoined", in something like the way that a root to an equation can be "adjoined" to a field. This is to be contrasted with the method that Gödel employed in 1938 to resolve the soundness problem for **AC.** Far from enlarging the universe of sets, Gödel *shrank* it, defining a subuniverse of *V* in which **AC** can be proved to hold. In doing so Gödel established the *relative consistency* of **AC** with respect to the standard axioms of set theory[1], namely that, if these latter are mutually consistent, then the addition of **AC** will leave that consistency undisturbed. It is interesting to note the similarity between the method used to prove the *consistency* of **AC** and that used in the 19th century to prove the *independence* of the parallel postulate (the method of "inner models"). In each case a model of the theory in question (set theory or geometry, respectively) augmented by the principle at issue (**AC** or the Bolyai-Lobachevsky postulate, respectively) is "carved out" from a "standard" model of the theory (the universe of sets *V* or Euclidean space, respectively).

Gödel's method of shrinking the universe of sets so as to obtain a model of **AC** rests on an essentially logical, or linguistic— as opposed to mathematical— idea, namely that of *definability.* He introduced a new hierarchy of sets—the *constructible* hierarchy —

---

[1] By that time, the standard axioms of set theory took the form of **ZF** (Zermelo-Fraenkel set theory with the axiom of foundation) or **VNB** (von Neumann- Bernays set theory).

by analogy with the cumulative type hierarchy. The latter is defined by the following recursion on the ordinals:

$$V_0 = \varnothing \qquad V_{\alpha+1} = \mathbf{P}V_\alpha \qquad V_\lambda = \bigcup_{\alpha<\lambda} V_\alpha \quad \text{for limit } \lambda$$

Here, for any set X, $\mathbf{P}X$ is the power set of X. The constructible hierarchy is defined by a similar recursion on the ordinals:

$$L_0 = \varnothing \qquad L_{\alpha+1} = \mathrm{Def}(L_\alpha) \qquad L_\lambda = \bigcup_{\alpha<\lambda} L_\alpha \quad \text{for limit } \lambda$$

In this case, for any set X, Def(X) is the set of all subsets of X which are first-order definable in the structure $(X, \in, (x)_{x\in X})$. The *constructible universe* is the class $L = \bigcup_{\alpha\in \mathrm{Ord}} L_\alpha$; the members of L are the *constructible sets*. Gödel showed that (assuming the axioms of Zermelo-Fraenkel set theory **ZF**) the structure (L, ∈) is a model of **ZF** and also of **AC** (as well as the Generalized Continuum Hypothesis). The relative consistency of **AC** with **ZF** follows[1].

It was also observed by Gödel[2] (and, independently, by others[3]) that a simpler proof of the relative consistency of **AC** can be formulated in terms of *ordinal definability.* If we write D(X) the set of all subsets of X which are first-order definable in the structure (X, ∈), then the class OD of *ordinal definable sets* is defined to be the union $\bigcup_{\alpha\in \mathrm{ORD}} D(V_\alpha)$. The class HOD of *hereditarily ordinal definable sets* consists of all sets *a* for which *a,* the members of *a,* the members of members of *a,* ... etc. are all ordinal definable. It can then be shown that the structure (HOD, ∈) is a model of **ZF** + **AC,** from which the relative consistency of **AC** with **ZF** again follows[4].

---

[1] For a detailed exposition of this proof, see Bell and Machover [1977].

[2] Gödel [1964].

[3] e.g. , Myhill and Scott [1971].

[4] For a detailed exposition, see Kunen [1980]. A compressed version is provided in Chapter IV.

CHRONOLOGY OF AC [1]

1904/1908. Zermelo introduces axioms of set theory, explicitly formulates **AC** and uses it to prove the well-ordering theorem, thereby raising a storm of controversy.
1904. Russell recognizes **AC** as the *Multiplicative Axiom*: the product of arbitrary nonzero cardinal numbers is nonzero.
1914. Hausdorff derives from **AC** the existence of nonmeasurable sets in the "paradoxical" form that ½ of a sphere is congruent to ⅓ of it[2].
1922. Fraenkel introduces the "permutation method" to establish independence of **AC** from a system of set theory with atoms[3].
1924. Building on the work of Hausdorff, Banach and Tarski derive from **AC** their *paradoxical decompositions of the sphere*: any solid sphere can be split into finitely many pieces which can be reassembled to form two solid spheres of the same size; and any solid sphere can be split into finitely many pieces in such a way as to enable them to be reassembled to form a solid sphere of arbitrary size.
1926. Hilbert introduces into his proof theory the "transfinite" or "epsilon" axiom as a version of **AC**[4].
1936. Lindenbaum and Mostowski extend and refine Fraenkel's permutation method.
1935-38. Gödel establishes the relative consistency of **AC** and the generalized continuum hypothesis with the standard axioms of set theory[5].

1 For a detailed history of the development of AC, see Moore [1982].
2 Hausdorff [1914],
3 Fraenkel [1922].
4 Hilbert [1926].
5 Gödel [1938], [1939], [1940].

1939-1954. In their famed work *Éléments de Mathématique,* Bourbaki adapts Hilbert's epsilon axiom so as to embed **AC** as a basic formal-logical principle[1] .

1963. Cohen proves the independence of **AC** and continuum hypothesis from the standard axioms of set theory[2].

[1] Bourbaki [1939]. Bourbaki employs the symbol "τ" in place of Hilbert's "ε". This may have been done to avoid typographical confusion with "∈", the basic symbol of set theory. Curiously, however, in first introducing a transfinite axiom, Hilbert used the symbol "τ", only there it was intended to represent the dual notion to that he later represented by "ε". Hilbert used the symbol "τ" to denote the operation of selecting an object which, if *it* happens to have a given property, then necessarily *every* object has that property. In his definitive later formulation of the transfinite axiom—that of the so-called "ε-calculus"—he used "ε" to denote the dual operation of selecting an object which, if *some* object happens to have a given property, then *it* necessarily has that property.

[2] Cohen [1963], [1963a], [1964].

II
Maximal Principles and Zorn's Lemma

THE NATURE AND ORIGINS OF MAXIMAL PRINCIPLES

AC is closely allied to a group of mathematical propositions collectively known as *maximal principles.* Broadly speaking, these propositions assert that certain conditions are sufficient to ensure that a partially ordered set (henceforth: *poset*) contains at least one *maximal element,* that is, an element such that, in the given partial ordering, no element strictly exceeds it.

To grasp the connection between the idea of a maximal element and **AC**, let us return to the latter's formulation **AC2** in terms of indexed sets. Thus suppose given an indexed family of nonempty sets $\mathcal{A} = \{A_i: i \in I\}$. Let us term a *partial choice function* on $\mathcal{A}$ any function f with domain $J \subseteq I$ such that $f(i) \in A_i$ for all $i \in J$. The set $\mathcal{F}$ of partial choice functions on $\mathcal{A}$ can be partially ordered by inclusion: we agree that, for $f, g \in \mathcal{F}, f \leq g$ provided that the domain of f is included in that of g and the value of f at an element of its domain coincides with the value of g there. It is now easy to see that each maximal element of P with respect to this partial ordering is a choice function on $\mathcal{A}$ (and conversely). For if m is a maximal element of $\mathcal{F}$, and the domain J of m fails to coincide with I, then there is $i_0 \in I$ such that $i_0 \notin J$. Now, choosing an arbitrary element a_0 of A_{i_0}, the set $m \cup \{<i_0, a_0>\}$ is a member of $\mathcal{F}$ properly including m, contradicting the latter's maximality. Accordingly the domain of m coincides with I and so m is a choice function on $\mathcal{A}$.[1]

[1] Notice, however, that this argument presupposes the correctness of the Law of Excluded Middle of classical logic. It does not go through if only intuitionistic logic is assumed. See Chapter VI below.

The existence of maximal elements yields a similar derivation of **AC3.** Thus suppose given a relation R with domain A and codomain B. Taking $\mathcal{F}$ to be the set of subfunctions of R, partially ordered by inclusion, one finds just as before that maximal elements of P are precisely the subfunctions of R with domain A.

Thus the existence of choice functions, and hence also **AC**, follows from the presence of maximal elements in sets of partial choice functions[1]. *Zorn's Lemma* is the best-known principle ensuring the existence of such maximal elements. To state it, we need a few definitions. Given a poset $(P, \leqslant)$, a subset C of P is called a *chain* in P if, for any $x, y \in C$, we have $x \leqslant y$ or $y \leqslant x$. An element m of P is *maximal* if, for all $x \in P$, $m \leqslant x$ implies $m = x$. P is said to be *inductive* if each chain in P has an upper bound in P. Zorn's Lemma may then be stated:

ZL *Any nonempty inductive poset has a maximal element*[2].

ZL can also be stated in an equivalent dual form. An element m of the poset P is *minimal* if, for all $x \in P$, $x \leqslant m$ implies $m = x$. P is said to be *reductive* if each chain in P has a lower bound in P. The dual form of Zorn's Lemma may then be stated:

DZL *Any nonempty reductive poset has a minimal element.*

ZL has an interesting history. In 1935 Zorn introduced it[3] as a "certain axiom on sets of sets" serving as a replacement for the "well-ordering theorem and its theory", which, he says, "are barred, from the algebraic point of view" in proving "the theorems of Steinitz concerning algebraic closure and the degree

---

[1] This holds only if classical logic is assumed. See previous footnote.
[2] Note that since the subset ∅ is a chain, an inductive set is always nonempty.
[3] Zorn [1935].

of transcendence". In so doing, he says, his purpose is "to make the proofs shorter and more algebraic". He seems to have been unaware of the fact that his principle had been previously given explicit formulation by Kuratowski in 1922[1], and even, in implicit form, by Hausdorff in 1909. Zorn refers to his principle as "a certain axiom" and later identifies it as "our *maximum principle*", so he presumably regarded it as less as a theorem (or lemma) than as a kind of postulate, on a par with **AC,** but superior to the latter in not requiring in its application the use of the cumbrous apparatus of ordinals and transfinite induction associated with the well-ordering theorem[2], which had come to be regarded by algebraists, particularly those of the Noether school, as "transcendental" devices, extraneous to the progress of mathematics. In the eyes of these mathematicians choice functions were no more than useful auxiliary devices, invested with no intrinsic mathematical significance. Thus it was natural that algebraists and other "working" mathematicians should come to prefer **ZL,** with its direct focus on maximality, to **AC**, given the fact that maximal objects had arisen naturally, and with striking frequency, within the abstract mathematics of the first half of the

---

[1] For this reason Zorn's Lemma is also known, particularly in Eastern Europe, as the "Kuratowski-Zorn" Lemma. While this is historically just, it is under the slick term "Zorn's Lemma" that the principle has entered the parlance of most mathematicians.

[2] The demonstration that every linear space has a basis using well-ordering, ordinals and transfinite induction provides a typical illustration of this cumbrousness. Thus suppose we are given a linear space L. Well-order L as $\{a_\xi: \xi < \alpha\}$ for some ordinal α. Using transfinite recursion define the sequence $<b_\xi>$ of elements of L as follows. First take $b_0 = a_0$. Then, for each ordinal $\xi > 0$, if $\{b_\eta: \eta < \xi\}$ does not generate L, let $b_\xi = a_\lambda$, where λ is the least ordinal such that a_λ is linearly independent of $\{b_\eta: \eta < \xi\}$. Otherwise let $b_\xi = a_0$. There must be an ordinal $\kappa < \alpha$ such that $\{b_\eta: \eta < \kappa\}$ generates L, for otherwise the map $\xi \mapsto b_\xi$ would be an injection of the class of all ordinals into L, in violation of the Axiom of Replacement. Let κ_0 be the least such κ. An argument using transfinite induction then shows that $\{b_\eta: \eta < \kappa_0\}$ is linearly independent and is therefore a basis for L.

20th century. Consider, for example, the fact that a basis of a linear space is simply a maximal independent subset; an algebraic closure of a field coincides with a maximal algebraic extension; a real closed field is a maximal real subfield of an algebraically closed field; a maximal ideal in a ring is just the kernel of an epimorphism to a field; a vertex of a convex figure is a minimal edge; a complete theory is a maximal consistent theory. There are very few analogous associations with choice functions[1], and none at all with well-orderings. It is therefore little wonder that **ZL** speedily replaced **AC** in the mathematicians' toolkit.

It is worth noting the fact that, unlike **AC**, **ZL** is still identified as a "Lemma" or a "Theorem", as opposed to an "Axiom". This suggests that **ZL** is, in the minds of mathematicians, a derivative principle, which, however useful and elegant it may be, still requires justification[2]. Its sole justification is, of course, **AC**. So it is of interest to see just how mathematicians have responded to the genuine challenge of presenting **ZL** as if it was a typical result of mathematics, straightforwardly provable without entanglement in the trappings of axiomatics. To quote from Serge Lang's influential book *Algebra*, **ZL** "could be just taken as an axiom of set theory". "However," he continues, "it is not psychologically satisfactory as an axiom, because its statement is too involved, and one does not visualize easily the existence of the maximal element asserted in the statement." The proof he then proceeds to give of **ZL** (based on the Bourbaki Fixed Point Lemma as stated and proved below) he describes as being based on "other properties of sets which

---

1 A few examples are provided in Chapter III.

2 Indeed a mathematical wag (Jerry Bona) has observed: "the Axiom of Choice is obviously true, the well-ordering theorem is obviously false, and, as for Zorn's Lemma, who can tell?

everyone would immediately grant as acceptable psychologically." Tellingly, in his proof he fails even to mention the use of **AC**! So it would seem that, along with Zermelo, Lang regards **AC** to be "acceptable psychologically"[1]. Bourbaki[2] goes even further in concealing the use of **AC**. Here, while remaining completely unmentioned (with the exception of a reference in the *Fascicule de Résultats* of the *Théorie des Ensembles*), **AC** is cleverly smuggled into the formal infrastructure of the *Élements de Mathématique* disguised as Hilbert's ε-symbol. By this means **AC** transcends mere psychological acceptability by simply vanishing into thin air!

AC is in fact easily derived from **ZL**. For the poset of partial choice functions, partially ordered by inclusion, on an indexed family of sets $\mathcal{A}$ is readily shown to be inductive; so, by the argument given at the beginning of the chapter, **ZL** yields the existence of a choice function on $\mathcal{A}$, that is, **AC2.** In a similar way, **ZL** yields **AC3**, in view of the fact that the set of subfunctions of a relation, partially ordered by inclusion, is also inductive.[3]

There is a less familiar way of deriving **AC** from **DZL** which echoes the "combinatorial" justification of **AC** sketched in Chapter I. Thus suppose given a family $\mathcal{S}$ of mutually disjoint nonempty sets; call a subset $S \subseteq \bigcup \mathcal{S}$ a *sampling* for $\mathcal{S}$ if, for any $X \in \mathcal{S}$, either $X \subseteq S$ or $S \cap X$ is nonempty and finite. Consider the set **S** of samplings, partially ordered by inclusion. Minimal elements of **S** – minimal samplings – are precisely the

---

[1] Which of course it is, indeed even objectively, at least for "pure" sets: see the final section of Chapter VI.

[2] Bourbaki [1939].

[3] The derivation of **AC** from **ZL** (but not the inductiveness of sets of partial choice functions) presupposes classical logic.

transversals for $\mathscr{S}$[1]; and the collection **S** of samplings is clearly nonempty since it contains $\bigcup\mathscr{S}$. So if it can be shown that **S** is reductive[2], Zorn's lemma will yield a minimal element of **S** and so a transversal for $\mathscr{S}$. The reductiveness of **S** can be seen as follows: suppose that $\{S_i : i \in I\}$ is a chain of samplings; let $S = \bigcap_{i\in I} S_i$. If we can show that S is itself a sampling, it will constitute a lower bound in **S** to $\{S_i : i \in I\}$. To this end let $X \in \mathscr{S}$ and suppose that $X \not\subseteq S$. Then there is $i \in I$ for which $X \not\subseteq S_i$; since S_i is a sampling, $S_i \cap X$ is finite nonempty, say $S_i \cap X = \{x_1, \ldots, x_n\}$. Clearly $S \cap X$ is then finite; suppose for the sake of contradiction that $S \cap X = \varnothing$. Then for each $k = 1, \ldots, n$ there is $i_k \in I$ for which $x_k \notin S_{i_k}$. It follows that $S_i \not\subseteq S_{i_k}$ for $k = 1, \ldots, n$, so, since the S_i form a chain, each S_{i_k} is a subset of S_i. Let S_j be the least of $S_{i_1}, \ldots, S_{i_k}$; then $S_j \subseteq S_i$ But since $x_k \notin S_j$ for $k = 1, \ldots, n$, it now follows that $S_j \cap X = \varnothing$, contradicting the fact that S_j is a sampling. Therefore $S \cap X \neq \varnothing$; and S is a sampling as claimed.

---

[1] That minimal samplings are transversals requires demonstration. Suppose S is a minimal sampling; then, given $X \in \mathbf{S}$, either (1) $S \cap X$ is finite nonempty or (2) $X \subseteq S$. In case (1) $S \cap X$ cannot contain two distinct elements because the removal of one of them from S would yield a sampling smaller than S, violating its minimality. So in this case $S \cap X$ must be a singleton. In case (2) B cannot contain two distinct elements a, b since, if it did, $S' = [(S - X) \cup \{a\}]$ would be a sampling smaller than S (notice that $S' \cap X = \{a\}$ and the relations of S' with the members of $\mathrm{S} - \{X\}$ are the same as those of S), again violating the minimality of S. So in this case X, and *a fortiori* $S \cap X$, must be a singleton.

[2] Notice that, had we elected to follow more closely the intuitive combinatorial derivation of **AC** as sketched in Chapter I by using cross-sections instead of samplings we would have encountered the obstacle that—unlike the set of samplings—the set of cross-sections is not necessarily reductive.

As we have seen, deriving **AC** from **ZL** is a comparatively straightforward matter[1]. The converse derivation, which serves to establish their equivalence[2], is considerably more laborious. Let us call a poset *strongly inductive* if each chain in it has a least upper bound.[3] We shall derive **ZL** from **AC** by first proving the

Bourbaki Fixed-Point Lemma.[4] Let $(P, \leqslant)$ be a strongly inductive poset , and let f be an inflationary self-map on P, i.e., a map $f: P \to P$ satisfying $x \leqslant f(x)$ for all $x \in P$. Then f has a fixed point.

Proof. Let us call a subset X of P *f-closed* if $f[X] \subseteq X$ and *f-inductive* if it contains the join (in P) of each of the f-closed chains it includes. Now fix some element $a \in P$ and let $\mathcal{K}$ be the collection of all subsets X of P satisfying the following conditions:

(i) $a \in X$;

(ii) $X \subseteq \{x \in P: a \leqslant x\}$;

(iii) X is f-closed;

(iv) X is f-inductive.

Since by hypothesis P itself satisfies these conditions, $\mathcal{K}$ is nonempty. Its intersection K is easily shown to satisfy (i) -(iv), and is accordingly the smallest subset of P to satisfy these conditions. We are going to show that K is a chain.

To establish this we define

$$K^* = \{x \in K: \forall y \in K[x \leqslant y \text{ or } f(y) \leqslant x]\},$$

1 Assuming classical logic.
2 Assuming classical logic.
3 When the poset is a family $\mathfrak{F}$ of sets partially ordered by inclusion, strong inductiveness is frequently established by showing that $\mathfrak{F}$ is *closed under unions of chains,* that is, the set-theoretical union of any chain in $\mathfrak{F}$ is again a member of $\mathfrak{F}$.
4 Bourbaki [1950].

and, for $b \in K^*$,

$$K_b = \{x \in K: x \leqslant b \text{ or } f(b) \leqslant x\}.$$

We first show that, for any $b \in K^*$, $K_b = K$. For this it suffices to show that K_b satisfies conditions (i) - (iv).

For condition (i) we observe that $b \in K$, so that $a \leqslant b$ since K satisfies (ii), and hence $a \in K_b$.

Condition (ii) follows immediately from the fact that K satisfies it.

To verify (iii), take $x \in K_b$. We have to show that $f(x) \in K_b$, i.e. $f(x) \in K$ and

(1) $$f(x) \leqslant b \text{ or } f(b) \leqslant f(x).$$

That $f(x) \in K$ follows from the fact that K is f-closed. To establish (1), note that since $b \in K^*$ and $x \in K$ we have

(2) $$b \leqslant x \text{ or } f(x) \leqslant b.$$

and since $x \in K_b$ we have

(3) $$x \leqslant b \text{ or } f(b) \leqslant x.$$

Taking the conjunction of (2) and (3) and using the distributive law of propositional logic, we get

(4) $$(b \leqslant x \text{ and } x \leqslant b) \text{ or } (b \leqslant x \text{ and } f(b) \leqslant x)$$
$$\text{or } (f(x) \leqslant b \text{ and } x \leqslant b)$$
$$\text{or } (f(x) \leqslant b \text{ and } f(b) \leqslant x).$$

The first disjunct of (4) gives $b = x$, so *a fortiori* $f(b) \leqslant f(x)$. The second gives $f(b) \leqslant x \leqslant f(x)$ since f is inflationary. The third and fourth both give $f(x) \leqslant b$. So (1) holds in all cases, and condition (iii) follows.

To establish (iv), let C be an f-chain in K_b. Then the join c of C is a member of K since the latter satisfies (iv). Since $C \subseteq K_b$,

$$\forall x \in C[x \leqslant b \text{ or } f(b) \leqslant x].$$

It follows that

$$(5) \qquad \forall x \in C(x \leqslant b) \text{ or } \exists x \in C\,(f(b) \leqslant x)$$ [1].

The first disjunct of (5) yields $c \leqslant b$, so that $c \in K_b$. The second disjunct gives $f(b) \leqslant c$; and so again $c \in K_b$. Condition (iv) follows.

Accordingly K_b satisfies (i) - (iv), so that $K_b = K$.

We next show in a similar way that $K^* = K$. Again it suffices to show that K^* satisfies conditions (i) - (iv).

Condition (i) follows immediately from the facts that $a \in K$ and K satisfies (ii).

Condition (ii) follows immediately from the fact that K satisfies it.

For condition (iii), suppose that $x \in K^*$. We need to show that $f(x) \in K^*$, i.e. $f(x) \in K$ and, for all $y \in K$,

$$(6) \qquad \forall y \in K[f(x) \leqslant y \text{ or } f(y) \leqslant f(x)].$$

That $f(x) \in K$ follows from the fact that K is f-closed. To establish (6), take $y \in K$. Then since $K_x = K$ (as shown above), $y \in K_x$, so we have

---

[1] As noted in Lawvwere and Rosebrugh [2003], This step uses the (intuitionistically invalid) logical law $\forall x\,[p(x) \vee q(x)] \Rightarrow \forall x p(x) \vee \exists x q(x)$, which is equivalent to the (also intuitionistically invalid) law $\forall x\,[p(x) \vee q] \Rightarrow \forall x p(x) \vee q$. Both of these may be seen as distributive laws. In Chapter V The latter of these is shown to be equivalent (over intuitionistic logic) to a choice rule.

(7) $$y \leqslant x \text{ or } f(x) \leqslant y.$$

Also $x \in K$, whence

(8) $$x \leqslant y \text{ or } f(y) \leqslant x.$$

(7) and (8) now yield (6) by means of an argument similar to the derivation of (1) from (2) and (3) above. Thus K^* satisfies condition (iii).

For condition (iv), let C be an f-chain in K^*. Then the join c of C is a member of K since the latter satisfies (iv). For each $x \in C$, $x \in K^*$, so $K_x = K$. Hence, for each $y \in K$, $y \in K_x$, so that, for all $x \in C$,

$$\forall x \in C[y \leqslant x \text{ or } f(x) \leqslant y]$$

It follows that, for each $y \in K$,

(9) $$\forall x \in C(f(x) \leqslant y) \text{ or } \exists x \in C(y \leqslant x).$$

The first disjunct of (9) yields $x \leqslant f(x) \leqslant y$ for every $x \in C$, whence $c \leqslant y$. If the second disjunct holds, then there is $x \in C$ for which $y \leqslant x$. Now $x \in K^*$, so either $f(y) \leqslant x$ or $x \leqslant y$. The first disjunct here gives $f(y) \leqslant c$; and from the second, conjoined with $y \leqslant x$, we infer that $y = x$. Hence $f(y) = f(x)$; but $f(x) \in C$ since C is an f-chain. It follows again that $f(y) \leqslant c$.

We have accordingly shown that, for every $y \in K$, either $c \leqslant y$ or $f(y) \leqslant c$, that is, $c \in K^*$. This establishes condition (iv).

To complete the proof that K is a chain, take $x, y \in K$. Then $x \in K^*$ and $y \in K_x$ since $K = K^* = K_x$. So $y \leqslant x$ or $f(x) \leqslant y$, whence $y \leqslant x$ or $x \leqslant y$ since f is inflationary. Hence K is a chain.

Since K is a chain and also satisfies (iii), it is an f-closed chain in K, and so has a join k which must be a member of K since K satisfies (iv). Also , since K is f-closed, $f(a)$ is a member of K. Therefore $f(k) \leqslant k$, and so, since f is inflationary, $f(k) = k$. Accordingly k is a fixed point of f, completing the proof of the lemma. ■

By the *Modified Zorn's Lemma* we shall mean the assertion

MZL. *Any strongly inductive poset has a maximal element.*

We can now use the Bourbaki fixed point lemma to derive **MZL** from **AC1.** For suppose given a strongly inductive poset $(P, \leqslant)$. Let g be a choice function for the family of sets $\{X \subseteq P: X \neq \emptyset\}$, and define $f: P \to P$ by

$$f(x) = x \text{ if } x \text{ is maximal in } P$$

$$f(x) = g(\{y \in P: x \leqslant y \text{ and } x \neq y\}) \text{ if } x \text{ is not maximal in } P.$$

Then f is inflationary and so by the Bourbaki lemma has a fixed point a; obviously a is a maximal element of P. ■

Finally we show that **ZL** follows from **MZL,** completing the demonstration of **ZL** from **AC.** In fact we shall show that both of these are equivalent to *Hausdorff's Maximal Principle:*

HMP *Every poset contains a maximal chain.*

Here by a maximal chain in a poset P we mean a chain in P which is maximal in the family of all chains in P.

Theorem. ZL, MZL *and* **HMP** *are all equivalent.*[1]

Proof. First note that **MZL** is an immediate consequence of **ZL**, so we need only prove the implications **MZL** ⇒ **HMP** and **HMP** ⇒ **ZL.**

[1] From the proofs given below, in which no use of the Law of Excluded Middle is made, it will be seen that these equivalences are constructively valid.

MZL ⇒ HMP. Let $(P, \leqslant)$ be a poset and let $\mathscr{C}$ be the family of all chains in P, partially ordered by inclusion. Then $\mathscr{C}$ is strongly inductive, since it is easily shown that the union of a chain (under $\subseteq$) of members of $\mathscr{C}$ is itself a member of $\mathscr{C}$. It follows now from **MZL** that $\mathscr{C}$ has a maximal element C; by definition C is a maximal chain in P.

HMP ⇒ ZL. Let $(P, \leqslant)$ be an inductive poset. Assuming **HMP**, P contains a maximal chain C. Since P is inductive, C has an upper bound c. We claim that c is a maximal element of P. For if $x \in P$ is such that $c \leqslant x$, then $C \cup \{x\}$ is a chain in P which includes C; the maximality of C implies that $C \cup \{x\} = C$, so that $x \in C$. Since c is an upper bound for C, it follows that $x \leqslant c$, whence $x = c$. So c is maximal, completing the proof.

CHRONOLOGY OF MAXIMAL PRINCIPLES[1].

1909. Felix Hausdorff introduces the first explicit formulation of a maximal principle (essentially **ZL**) and derives it from **AC.**

1914. Hausdorff's *Grundzüge der Mengenlehre* (one of the first books on set theory and general topology) includes a number of maximal principles, including what we have called **HMP.**

1922. Kazimir Kuratowski formulates and employs several maximal principles, including ZL.

1926-28. Salomon Bochner and others independently introduce maximal principles.

1935. Max Zorn, seemingly unacquainted with previous formulations of maximal principles, publishes his definitive

[1] For a detailed history of maximal principles, see Moore [1982].

version thereof later to become celebrated as **ZL.** First formulated in Hamburg in 1933, **ZL** as quickly "adopted" by Claude Chevalley and Emil Artin. It seems to have been Artin who first recognized that **ZL** would yield **AC**, so that the two are set-theoretically equivalent.

1939-40. Teichmüller, Bourbaki and Tukey independently reformulate **ZL** in terms of "properties of finite character". If *A* is a set, and *P* a property of subsets of *A* (in this case we shall say that *P* is *A-based*), then *P* is said to be of *finite character* if, for any subset *X* of *A*, *X* has *P* if and only if every finite subset of *X* has *P*. Then **ZL** is equivalent to the assertion that, for any set *A*, and any *A*-based property *P* of finite character, there is a maximal subset of *A* possessing *P*.

III
Mathematical Applications of the Axiom of Choice

When Zermelo introduced **AC** he recognized its fundamental nature, and so also grasped its potential significance for the development of mathematics. But even Zermelo himself could not have anticipated the extraordinary wealth of mathematical propositions whose demonstrations ultimately depend on his principle, many of which have turned out to be formally equivalent to it[1]. It is of interest to note that by the 1930s mathematicians had come to realize that the simplest and most direct way of deriving the majority of such propositions is in fact *not* to employ **AC** *per se,* but rather to use a maximal principle such as **ZL**. Indeed, as already pointed out, Zorn introduced his Lemma precisely so as to avoid the use of the well-ordering theorem, with the attendant apparatus of ordinals and transfinite induction whose use was often required when applying **AC**.

In this chapter we list, and, where it seems appropriate, sketch proofs of, a number of propositions, from a range of areas of mathematics, whose demonstrations require the use of **AC** or **ZL.** We begin with those propositions—call them **AC**-propositions—whose simplest demonstrations employ **AC** (or the well-ordering theorem) and then turn to the considerably lengthier list of propositions—**ZL**-propositions—which are much more directly proved by using **ZL.**

---

[1] For equivalents of **AC**, see Rubin and Rubin [1985]; for consequences, see Howard and Rubin [1998].

AC-PROPOSITIONS

- **The Multiplicative Axiom**[1] **The product of any set of non-zero cardinal numbers is non-zero.** This is equivalent to **AC.**
- **Each infinite set has a denumerable subset.** This amounts to showing that, for any infinite set A, there is an injective map ω [2]$\rightarrow A$. Using **AC1**, let f be a choice function on the family of non-empty subsets of A (note that since A is infinite, it is itself non-empty. Now define the map $g\colon \omega \rightarrow A$ by recursion as follows: $g(0) = f(A)$, $g(n+1) = f(A - \{g(0), \ldots, g(n)\})$. Then g is an injection of ω into A.
- **The equivalence of various definitions of finiteness.** These include: (1) a set is finite provided it is equipollent to a set of the form $\{0, \ldots, n\}$; (2) *Dedekind-Peirce finiteness*: a set A is *DP*-finite iff every injection $A \rightarrow A$ is surjective; (3) *Kuratowski-finiteness*: a set is *K*-finite iff it is a member of the least class $\mathcal{K}$ of sets that contains $\varnothing$ and all singletons, and is closed under unions of pairs of its members; (4) *Tarski-finiteness:* a set A is T-finite iff every total ordering on it is a well-ordering.
- **The Principle of Dependent Choices**[3] **.** ***For any nonempty relation R on a set A for which*** range **(*R*)** $\subseteq$ domain**(*R*)*, there is a function*** $g : \omega \rightarrow A$ ***such that, for all*** $n \in \omega, R(g(n), g(n+1))$. To prove this, again let f be a choice function on the family of non-empty subsets of A, and let a be some element of range(R). Now define the map $g\colon \omega \rightarrow A$ by recursion as follows: $g(0) = a$, $g(n + 1) = f(\{x\colon (g(n), x) \in R\}$. Then g satisfies the required conditions.

---

[1] Russell [1906].

[2] As is customary, we use ω to denote the set of natural numbers. It is hoped that this will not cause confusion with other uses of ω in this book, notably in Chapter VII.

[3] Bernays [1942], Tarski [1948].

- **Distributive laws: for any doubly indexed family of sets** $\{A_{ij}: \langle i,j\rangle \in I \times J\}$,

$$\bigcap_{i\in I}\bigcup_{j\in J} A_{ij} = \bigcup_{f\in J^I}\bigcap_{i\in I} A_{if(i)} \qquad \prod_{i\in I}\bigcup_{j\in J} A_{ij} = \bigcup_{f\in J^I}\prod_{i\in I} A_{if(i)}.$$

It is not difficult to show that both of these are equivalent to **AC1.**[1]

- **Existence of a Lebesgue non-measurable set of real numbers**[2]**.** To indicate how it is derived from **CAC**, let E be the equivalence relation on the interval (0, 1) defined by xEy iff $x - y$ is rational, and, using **CAC**, let $A \subseteq (0, 1)$ be a transversal for the family of E-equivalence classes. It is then not hard to show that A is non-measurable[3].

- **Projectivity of sets and freely generated objects.** An object E of a category $\mathscr{C}$ is *projective* if the diagram (with f epi)

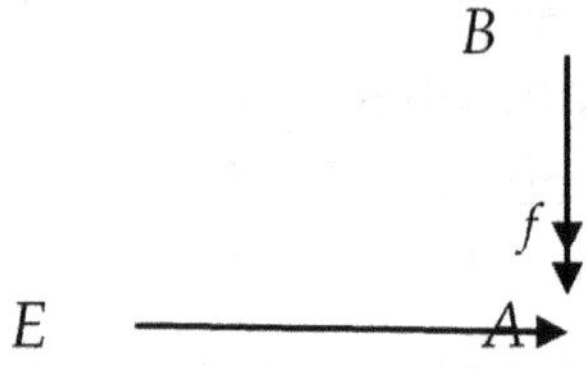

can be completed to a commutative diagram

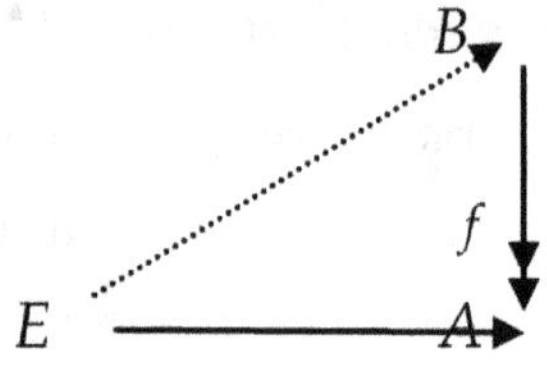

[1] As also are the corresponding assertions with "=" replaced by "⊆".

[2] Vitali [1905]. This was shown much later to be a consequence of **BPI** (see below) and hence weaker than **AC.** Solovay [1970] established its independence of the axioms of set theory.

[3] For a full proof of this see, e.g. Kestelman [1960].

An object E of $\mathscr{C}$[1] is *freely generated* by a set I (or simply *free on* I) if $I \subseteq E$ and, for any object A of $\mathscr{C}$, each map $f: I \to A$ is uniquely extensible to a $\mathscr{C}$- arrow $E \to A$. It is easily shown that **AC1** is equivalent to the assertion that **every set is projective**. **AC1** also implies that **every free Abelian group, and every free Boolean algebra, is projective**[2]. To prove the first assertion (the proof of the second being similar) , suppose that E is an Abelian group free on the set I, and that $f: E \to A$ and $g:B \twoheadrightarrow A$ are morphisms, with g epi. Since g is epi, **AC1** gives a map $k: I \to B$ for which $k(i) \in g^{-1}(f(i))$ for all $i \in I$, from which it follows that $g \circ k = f$. Since E is free on I, k extends uniquely to a morphism $h: E \to B$. Then $g \circ h = f$ because both have the same restrictions to I.

- **Neilsen-Schreier Theorem: each subgroup of a free group is free.** This is usually proved by means of the well-ordering theorem, but it can also be proved using **ZL.** The details are, however, too involved to be presented here.

- **Łoś's Theorem.** For each $i \in I$ let $\mathfrak{A}_i$ be a relational structure $<A_i, R_i>$ with R_i a binary relation on A_i. If U is an ultrafilter (i.e., a maximal proper filter) in **P***I*, define the relation $\approx_F$ on $\prod_{i \in I} A_i$ by $f \approx_F g$ iff $\{i \in I: f(i) = g(i)\} \in U$. It is easily shown that $\approx_F$ is an equivalence relation on $\prod_{i \in I} A_i$. For each $f \in \prod_{i \in I} A_i$ write f_U for the $\approx_F$ -equivalence class of f, and let $\prod_{i \in I} A_i / U = \{f_U : f \in \prod_{i \in I} A_i\}$. Define the relation R_U on $\prod_{i \in I} A_i$ by stipulating that $<f_U, g_U> \in R_U$

[1] Here we assume that the category $\mathscr{C}$ is concrete in that the sense that its objects is a set with additional structure and its arrows are mappings in the set-theoretical sense.

[2] The projectivity of free Abelian groups was proved equivalent to **AC** by Blass [1979].

iff $\{i \in I: \langle f(i), g(i)\rangle \in R_i\} \in U$. Finally we define the *ultraproduct* $\prod_{i \in I} \mathfrak{A}_i / U$ of the family of structures $\{\mathfrak{A}_i: i \in I\}$ over U to be the structure $\langle \prod_{i \in I} A_i, R_U\rangle$. If each $\mathfrak{A}_i$ is identical with a fixed structure $\mathfrak{A}$, $\prod_{i \in I} \mathfrak{A}_i / U$ is called an *ultrapower* of $\mathfrak{A}$ and is written $\mathfrak{A}^I/U$.

Łoś's Theorem asserts that, for any formula $\varphi(x_1, \ldots, x_n)$ of the first-order language for binary relational structures, and any $f_1, \ldots, f_n \in \prod_{i \in I} A_i$,

$$\prod_{i \in I} \mathfrak{A}_i / U \models \varphi[f_1/F, \ldots, f_n/F] \textit{ iff } \{i \in I: \mathfrak{A}_i \models \varphi[f_1(i), \ldots, f_n(i)]\} \in U$$ [1].

As an immediate consequence, for any sentence σ,

(*) $$\mathfrak{A}^I/U \models \sigma \textit{ iff } \mathfrak{A} \models \sigma$$ [2].

The theorem is proved by induction on the logical symbols in φ; **AC** is invoked in the case in which φ is the form $\exists x \psi(x)$ [3].

The implication from **AC** to Łoś's Theorem cannot be reversed. For a model $\mathfrak{M}$ of **ZF** has been constructed[4] within which every ultrafilter in a power set is principal, that is, generated by a singleton. In $\mathfrak{M}$, **AC** fails but Łoś's Theorem holds trivially since $\mathfrak{A}^I/U$ is isomorphic to $\mathfrak{A}$ whenever U is principal.

On the other hand **AC** can be derived from (and so is equivalent to) Łoś's Theorem + the Boolean Prime Ideal Theorem **BPI**[5]. This new proof seems sufficiently neat to merit presentation. We shall actually derive **AC3*** from **(*)** above + **BPI.** First,

---

[1] Here we write $\mathfrak{A} \models \varphi[a_1, \ldots, a_n]$ for "$a_1, \ldots, a_n$ satisfies the formula φ in the structure $\mathfrak{A}$".

[2] Here we write $\mathfrak{A} \models \sigma$ for "the sentence σ holds in the structure $\mathfrak{A}$". In model-theoretic jargon, **(*)** asserts that $\mathfrak{A}$ and $\mathfrak{A}^I/U$ are *elementarily equivalent.*

[3] See, e.g. Bell and Slomson [2006].

[4] Blass [1977]

[5] Originally proved by Howard [1975]. For the **BPI** see below.

assuming (*), we prove the following. Let $R \subseteq A \times A$ satisfy $\forall x{\in}A\exists y{\in}A\ xRy$. For $f \in A^A$ let $S(f) = \{x \in A: xRf(x)\}$. Then:

(**) *for any ultrafilter U in* **P***A, there is* $f \in A^A$ *for which* $S(f) \in U$.

To prove (**), let $\mathfrak{A}$ be the structure $\langle A, R\rangle$ and let U be an ultrafilter in **P***A***.** Then $\mathfrak{A} \models \forall x\exists y\ xRy$, so by (*) $\mathfrak{A}^I/U \models \forall x\exists y\ xRy$. It follows that, if we write i for the identity function on A, there is $f \in A^A$ for which $i/U\ R_U\ f/U$, whence $S(f) = \{x\in A: i(x)\ R\ f(x)\} \in U$.

To derive **AC3*** we need to show that there is $f \in A^A$ for which $S(f) = A$. We note first that, for any $f, g \in A^A$, $S(f) \cup S(g) = S(h)$, where $h \in A^A$ is defined by

$$h(i) = f(i) \ \text{ if } i \in S(f) \text{ or } i \notin S(g)$$
$$h(i) = g(i) \ \text{ if } i \in S(g) \text{ and } i \notin S(f).$$

Now suppose for contradiction's sake that $S(f) \neq A$ for all $f \in A^A$. Then, using what we have just noted, the ideal in **P***I* generated by $\{S(f): f \in A^A\}$ is proper and so, by **BPI**, included in a maximal ideal M. Then $U = \{A\backslash X: X \in M\}$ is an ultrafilter in **P***I* not containing any $S(f)$, in contradiction with (**). Thus $S(f) = A$ for some $f \in A^A$. **AC3*** follows.

- **Löwenheim - Skolem - Tarski Theorem[1] — a first-order sentence having a model of cardinality κ also has a model of cardinality μ provided $\aleph_0 \leq \mu \leq \kappa$.** This was proved equivalent to **AC** by Tarski.

[1] Löwenheim [1915], Skolem [1920], Tarski and Vaught [1957].

ZL-PROPOSITIONS

- **The Well-Ordering Theorem**[1]**: every set can be well-ordered.** This is equivalent to **AC.** While Zermelo's original proof of this was based on **AC**, the use of **ZL** furnishes a much more efficient proof. Thus let $\mathscr{B}$ be the set of all pairs $\langle B, \leq \rangle$ with $B \subseteq A$ and $\leq$ a well-ordering of B. Then $\mathscr{B} \neq \varnothing$; partially order $\mathscr{B}$ by

$$\langle B, \leq \rangle \sqsubseteq \langle B', \leq' \rangle \Leftrightarrow B \subseteq B', \leq \textit{ is the restriction of } \leq' \textit{ to } B, \textit{ and } B \textit{ is an initial segment of } B'.$$

A straightforward argument shows that $\langle \mathscr{B}, \sqsubseteq \rangle$ is closed under unions of chains, hence (strongly) inductive. Consequently, **ZL** applies to furnish a maximal element $\langle D, \leq \rangle$. This maximal element is easily shown to be a well-ordering of A.

- **The Ordinal Covering Principle**: **for each set X, there is a surjection from an ordinal onto X.** This is an immediate consequence of the well-ordering theorem and is easily seen to be equivalent to **AC.**

- **The Order Extension Principle: every partial ordering on a set can be extended to a total ordering.** Let $(P, \leq)$ be a poset and consider the set $\mathscr{R}$ of all partial orderings on P extending $\leq$. It is easy to show that $\mathscr{R}$ is closed under unions of chains and so by **ZL** has a maximal element R. We shall show that R is a linear ordering extending $\leq$. For this it suffices to show that, for any $p, q \in P$, either pRq or qRp. Suppose on the contrary that there exist $p, q \in P$ such that $\neg pRq \wedge \neg qRp$. Let S be the relation $\{(x,y): xRp \wedge qRy\}$ and R' the relation $R \cup S$. We show that R' is a

[1] Zermelo [1904], [1908].

linear ordering of P, contradicting the maximality of R since $R \subseteq R'$and $R \neq R'$.

Clearly R' is symmetric. To show that it is transitive, suppose that $xR'y$ and $yR'z$. Then one of the following cases holds:

(i) $xRy \wedge yRz$;

(ii) $xRy \wedge yRp \wedge qRz$;

(iii) $xRp \wedge qRy \wedge yRz$;

(iv) $xRp \wedge qRy \wedge yRp \wedge qRz$.

Case (iv) is impossible, since it implies qRp, which contradicts the hypothesis. In case (i) the transitivity of R gives xRz, and so also $xR'z$. In cases (ii) and (iii) the transitivity of R gives $xRp \wedge qRz$, i.e. xSz, and consequently $xR'z$. This shows that $R'$is transitive.

To show that $R'$is antisymmetric, suppose that $xR'y$ and $yR'x$. We then have 4 cases analogous to (i) – (iv) where z is replaced by x. Cases (ii) – (iv) are impossible, and case (i) implies $x = y$. This proves the Order Extension Principle.

- **The Comparability Principle for cardinal numbers: for any cardinal numbers $\mathfrak{m}$, $\mathfrak{n}$, either $\mathfrak{m} \leqslant \mathfrak{n}$ or $\mathfrak{n} \leqslant \mathfrak{m}$.**[1] This amounts to showing that, for any pair of sets A, B, there is an injection of one into the other. This is efficiently proved by means of **ZL**. For consider the set $\mathscr{F}$ of all injective maps of subsets of A into B, partially ordered by inclusion. It is readily shown that $\mathscr{F}$ is closed under unions of chains, and accordingly (strongly) inductive. **ZL** then supplies a maximal element F of $\mathscr{F}$; it is not difficult to show that either domain(F) = A or range(F) = B. In the first case, we have an injection of A into B; in the second case, vice-versa.

[1] The comparability principle was shown to be equivalent to **AC** by Hartogs [1915].

- **The Idempotency Principle for infinite cardinal numbers: each infinite cardinal number is equal to its square.**[1] This is most efficiently derived from **ZL**. For the proof, in addition to **ZL**, one uses the already established fact that each infinite set has a denumerable subset; the elementary set theoretic facts that $\omega \times \omega \approx^{2} \omega$; $k\mathfrak{m} \leq \mathfrak{m}^2$ for any infinite cardinal m and any natural number k; and finally the *Schröder-Bernstein theorem* (whose proof does not require **AC**) that, for any cardinals $\mathfrak{m}$, $\mathfrak{n}$, if $\mathfrak{m} \leqslant \mathfrak{n}$ and $\mathfrak{n} \leqslant \mathfrak{m}$, then $\mathfrak{m} = \mathfrak{n}$.

Accordingly let m be an infinite cardinal; we show that $\mathfrak{m}^2 = \mathfrak{m}$. Let $\mathfrak{m} = |A|$[3] and let $B \subseteq A$ satisfy $B \approx \omega$. Then there is a bijection f_0: $B \to B \times B$. Let $\mathscr{F}$ be the set of pairs $\langle X, f\rangle$ where $B \subseteq X \subseteq A$ and f is a bijection between X and $X \times X$ such that $f_0 \subseteq f$. Partially order $\mathscr{F}$ by stipulating that

$$\langle X, f\rangle \sqsubseteq \langle X', f'\rangle \Leftrightarrow X \subseteq X' \text{ and } f \subseteq f'.$$[4]

Then $\langle \mathscr{F}, \sqsubseteq\rangle$ is easily shown to be (strongly) inductive and hence by **ZL** has a maximal element $\langle C, g\rangle$. We show that $|C| = \mathfrak{m}$; since $C \approx C \times C$, it will follow that $\mathfrak{m} = \mathfrak{m}^2$.

Suppose on the contrary that $|C| < \mathfrak{m}$. Then since $\mathfrak{n} = |C|$ is infinite and $\mathfrak{n}^2 = \mathfrak{n}$ (recall that $C \approx C \times C$), we have

$$\mathfrak{n} \leqslant 2\mathfrak{n} \leqslant 3\mathfrak{n} \leqslant \mathfrak{n}^2 = \mathfrak{n}.$$

---

[1] This was originally proved by Hessenberg [1906] and shown to be equivalent to **AC** by Tarski [1924]. The proof given here, based on that of Zorn [1944], is Bourbaki's [1963].
[2] For sets A, B we write $A \approx B$ to assert the existence of a bijection between A and B, so that A and B have the same cardinality.
[3] We use $|A|$ to denote the cardinality of A.
[4] In the sequel we shall call such a partial ordering a *partial ordering by extension.*

It follows from the Schröder-Bernstein theorem that $3\mathfrak{n} = 2\mathfrak{n} = \mathfrak{n}$. From $\mathfrak{n} < \mathfrak{m}$ we infer that $|A \setminus C| > \mathfrak{n}$; for, if not, then

$$|A| \leqslant \mathfrak{n} + \mathfrak{n} = 2\mathfrak{n} = \mathfrak{n},$$

contradicting $|A| > |C| = \mathfrak{n}$. Accordingly there is a subset $Y \subseteq A \setminus C$ such that $|Y| = \mathfrak{n}$; put $Z = C \cup Y$. We show that there is a bijection $h: Z \to Z \times Z$ such that $g \subseteq h$. For we have

$$Z \times Z = (C \times C) \cup (C \times Y) \cup (Y \times C) \cup (Y \times Y),$$

and the sets on the right hand side of this equality are disjoint. Since $C \approx Y$, we have

$$|C \times Y| = |Y \times C| = |Y \times Y| = \mathfrak{n}^2 = \mathfrak{n},$$

so that

$$|(C \times Y) \cup (Y \times C) \cup (Y \times Y)| = 3\mathfrak{n} = \mathfrak{n}.$$

Thus there is a bijection g' of Y onto $(C \times Y) \cup (Y \times C) \cup (Y \times Y)$. Now let h be the unique map h of Z into $Z \times Z$ whose restriction to C is g and whose restriction to Y is g'. Then h is a bijection and $g \subseteq h$. But this contradicts the maximality of $<C, g>$. Therefore $|C| < \mathfrak{m}$ is impossible, and so, since evidently $|C| \leqslant \mathfrak{m}$, it follows that $|C| = \mathfrak{m}$ and the result is proved.

- **The Boolean Prime Ideal Theorem (BPI).** This is the assertion that every Boolean algebra contains a prime ideal, or equivalently, a prime filter. It is proved by showing that any maximal ideal in a Boolean algebra is prime, and then invoking the fact above that every distributive lattice, and *a fortiori* every Boolean algebra, has a maximal ideal. **BPI** was shown to be weaker than **AC** by Halpern and Levy [1971]. **BPI** is equivalent to the **Boolean Ultrafilter Theorem** which asserts that every Boolean algebra contains an ultrafilter, that is, a maximal filter. This can be strengthened to the assertion that, for any Boolean algebra *B*, any subset X with the finite meet property is contained in an ultrafilter. Here X has the *finite meet property* if the meet of any finite subset of X is $\neq 0$.[1]

- **The Stone Representation Theorem for Boolean algebras—every Boolean algebra is isomorphic to a field of sets.**[2] This is proved by considering, for a given Boolean algebra B, the set $S(B)$ of all prime ideals of B, and the map $u: B \to \mathbf{P}S(B)$ defined by $u(x) = \{I \in S(B): x \notin I\}$. Then u is an epimorphism of B onto the field of sets $\{u(x): x \in B\}$, and injective —hence an isomorphism— because of **BPI.**
- **The Sikorski Extension Theorem for Boolean algebras—every complete Boolean algebra is injective**[3]. A Boolean algebra C is *injective* if, for any Boolean algebra B, and any subalgebra A of B, any morphism $A \to C$ can be extended to a morphism $B \to C$. To prove the theorem using **ZL**, let C be a complete Boolean algebra, A a subalgebra of B, and $h: A \to C$ a morphism. Let $\mathcal{F}$ be the set of pairs $\langle D, f\rangle$, where D is a subalgebra of B containing A, and $f: D \to C$ is a morphism extending h. Then $\mathcal{F}$, partially ordered by extension, is (strongly) inductive, and so by **ZL** has a maximal element $\langle M, g\rangle$. We show that $M = B$, from which the theorem immediately follows.

Let b be an arbitrary element of B, and let c be any element of C satisfying

$$\bigvee\{g(x): x \leqslant b \;\&\; x \in M\} \leqslant c \leqslant \bigwedge\{g(y): b \leqslant y \;\&\; y \in M\}.$$

Let M_b be the subalgebra of B generated by $M \cup \{b\}$. Each element u of M_b can be expressed in the form

[1] When the Boolean algebra in question is a field of sets, the finite meet property is referred to as the finite *intersection* property.

[2] This was first proved by Stone [1936]. It is equivalent to the **BPI** and hence weaker than **AC.**

[3] Sikorski [1948].

$u = (x_u \wedge b) \vee (y_u \wedge b^*)$ with $x, y \in M$. If we now define $g' : M_b \to C$ by $g'(u) = (g(x_u) \wedge b) \vee (g(y_u) \wedge b^*)$, then g' is a (well-defined) morphism extending g, and so $<M_b, g'>$ is an extension of $<M, g>$ in $\mathscr{F}$. From the maximality of $<M, g>$ it follows that $M_b \subseteq M$, so that $b \in M$.

Since b was arbitrary, we conclude that $M = B$, and the theorem follows[1].

The question of the equivalence of this theorem with **AC** is one of the few remaining open questions of interest in this area; it was proved independent of **BPI** by Bell [1983][2]. In Bell [1988a] the Sikorski Extension Theorem was shown to be equivalent to the following strengthening of **BPI**: for any Boolean algebra A and any subalgebra B of A, there is an ideal I in A maximal with respect to the property $I \cap B = \{0\}$.

- **The Disjointness Principle for complete Boolean algebras: choice functions as elements of Boolean fuzzy sets.** The *Disjointness Principle for complete Boolean algebras* is the following assertion. Let B be a complete Boolean algebra, I a set and $\{a_i: i \in I\}$ an I-indexed subset of B satisfying $\bigvee_{i \in I} a_i = 1$[3]. Then there exists an I-indexed subset $\{b_i: i \in I\}$ of B such that (1) $b_i \leq a_i$

[1] Observe that, unlike the majority of the derivations from **ZL**, no use of the Law of Excluded Middle has been made in this instance. In fact, the Sikorski extension theorem is (as far as I know) one of the very few results of significance constructively derivable from **ZL**.

[2] There it is shown, *inter alia*, that the Sikorski Extension Theorem holds in a model $\mathfrak{M}$ of set theory iff **BPI** holds in every Boolean extension of $\mathfrak{M}$. Accordingly to prove the independence of **AC** from the Sikorski Extension Theorem it would appear to be necessary to construct a model $\mathfrak{M}$ of set theory in which **AC** fails but in every Boolean extension of $\mathfrak{M}$ **BPI** holds. A daunting task indeed.

[3] If X is a subset of a partially ordered set P, $\bigvee X$ and $\bigwedge X$ denote, respectively, the join or least upper bound and the meet or greatest lower bound, respectively, of X, assuming these exist.

for all $i \in I$; (2) $b_i \wedge b_j = 0$ for $i \neq j$; (3) $\bigvee_{i \in I} b_i = 1$. This can be proved using **ZL** as follows. Let $\mathscr{D}$ be the collection of all I-indexed subsets $X = \{x_i: i \in I\}$ of B such that (i) $x_i \leq a_i$ for all $i \in I$; (ii) $x_i \wedge x_j = 0$ for $i \neq j$. Partially order $\mathscr{D}$ by decreeing that $X \sqsubseteq Y$ iff $\forall i\ (x_i \leq y_i)$. Then $\mathscr{D}$ is (strongly) inductive. For suppose $\mathscr{X} = \{X_k: k \in K\}$ is a chain in $\mathscr{D}$ with $X_k = \{x_{ik}: i \in I\}$ for each $k \in K$. For each $i \in I$ let $c_i = \bigvee_{k \in K} x_{ik}$ and let $C = \{c_i: i \in I\}$. We show that $C \in \mathscr{D}$. Since each $x_{ik} \leq a_i$, $c_i \leq a_i$, i.e. C satisfies (i). To show that C satisfies (ii), let $i \neq j$ and note that

$$(*) \qquad c_i \wedge c_j = \bigvee_{k \in K} x_{ik} \wedge \bigvee_{k' \in K} x_{jk'} = \bigvee_{k \in K} \bigvee_{k' \in K} x_{ik} \wedge x_{jk'}.$$

Now since $\mathscr{X}$ is a chain either $\forall i\ (x_{ik} \leq x_{ik'})$ or $\forall i\ (x_{ik'} \leq x_{ik})$. In the first case $x_{ik} \wedge x_{jk'} \leq x_{ik'} \wedge x_{jk'} = 0$, and similarly in the second case. So in either case $c_i \wedge c_j = 0$ follows from (*), which shows that C satisfies (ii). So $C \in \mathscr{D}$. Since C is obviously a (least) upper bound for $\mathscr{X}$, it follows that $\mathscr{D}$ is inductive. Therefore by **ZL** $\mathscr{D}$ has a maximal element $M = \{b_i: i \in I\}$. Clearly M satisfies (1) and (2); it remains to show that it satisfies (3). Suppose not; then $d = \bigwedge_{i \in I} b_i^* \neq 0$. Since $\bigvee_{i \in I} a_i = 1$, it follows that $0 \neq d = d \wedge \bigvee_{i \in I} a_i = \bigvee_{i \in I} d \wedge a_i$. Therefore $d \wedge a_{i_0} \neq 0$ for some i_0. Now define b'_i by $b'_{i_0} = b_{i_0} \vee (d \wedge a_{i_0})$, $b'_i = b_i$ for $i \neq i_0$ and let $M' = \{b'_i: i \in I\}$. It is easily verified that $M' \in \mathscr{D}$, $M' \neq M$, and $M \sqsubseteq M'$. This contradicts the maximality of M, and we conclude that M satisfies (3). The disjointness principle is accordingly proved.

The disjointness principle has a natural formulation in terms of *Boolean fuzzy sets.* Given a complete Boolean algebra B, a *B-fuzzy set* is a pair $\mathbf{I} = (I, a)$ consisting of a set I and a map $a\colon I \to B$. We write a_i for $a(i)$; think of a_i as the "Boolean truth value" of the statement $i \in \mathbf{I}$. A *fuzzy map* $p\colon (I, a) \to (J, b)$ is a map $p\colon I \times J \to B$ such that (1) $\forall i{\in}I\ \forall j{\in}J\ (p_{ij} \leqslant b_j)$; (2) $p_{ij} \wedge p_{ij'} = 0$ if $j \neq j'$; (3) $\bigvee_{j\in J} p_{ij} = a_i$. Here we think of p_{ij} as the "Boolean truth value" of the statement $p(i) = j$; in that case (1), (2) and (3) are the Boolean versions of, respectively, range $(p) \subseteq \mathbf{J}$; p is single valued; and domain $(p) = \mathbf{I}$. B-fuzzy sets and maps comprise the objects and arrows of a category $\mathcal{F}uz_B$ in which the identity arrow on $\mathbf{I} = (I, a)$ is the map $1_\mathbf{I}\colon I \times I \to B$ defined by $1_\mathbf{I}(i, i) = a_i$ and $1_\mathbf{I}(i, i') = 0$ if $i \neq i'$; while the composite qp of two arrows $p\colon (I, a) \to (J, b)$ and $q\colon (J, b) \to (K, c)$ is given by $(qp)_{ik} = \bigvee_{j\in J} p_{ij} \wedge q_{jk}$. $\mathcal{F}uz_B$ has a terminal object $\mathbf{1} = (\{0\}, u)$ with $u\colon \{0\} \to B$ the map with value 1.

It is now easily shown that an arrow $1 \to \mathbf{I} = (I, a)$ in $\mathcal{F}uz_B$, that is, an *element* of $\mathbf{I}$ in $\mathcal{F}uz_B$ corresponds precisely to an I-indexed subset $\{b_i\colon i \in I\}$ of B satisfying conditions (1) – (3) above: let us call such a *Boolean element* of $\mathbf{I}$. Moreover, the Boolean formulation of the condition "$\mathbf{I}$ is nonempty" is $\bigvee_{i\in I} a_i = 1$. Accordingly the disjointness principle may be translated as *any nonempty Boolean fuzzy set has a Boolean element.* Let us call this latter the *Boolean element principle (***BEP**).

Since the disjointness principle is a consequence of **ZL**, and hence of **AC**, so is **BEP.** In fact these are all equivalent. It suffices then to show that **BEP** implies **AC**. A nice way of doing this is to show that, for any indexed family of nonempty sets $\mathcal{A} = \{A_j\colon j \in J\}$, choice functions on $\mathcal{A}$ correspond precisely to Boolean elements of a certain nonempty Boolean fuzzy set. In fact, let B be the complete Boolean algebra **P**J of all subsets of J and $I = \{<j,x>: x \in A_j\}$. Now define $\sigma: I \to B$ by $\sigma(<j,x>) = \{j\}$. Then $\bigcup_{i \in I} \sigma_i = J = 1$ in B, so (I, σ) is a nonempty B-fuzzy set.

We now show that Boolean elements of (I, σ) correspond to choice functions on $\mathcal{A}$. In fact, each Boolean element corresponds to an I- indexed family $\{P_i: i \in I\}$ for which $P_i \subseteq \sigma(i)$ for $i \in I$, $P_i \cap P_{i'} = \varnothing$ for $i \neq i'$, and $\bigcup_{i \in I} P_i = J$. These in turn correspond to choice functions on $\mathcal{A}$. For if $\{P_i: i \in I\}$ is such a family, there is for each $j \in J$ a unique $i \in I$ for which $j \in P_i$. Then $i = <j', x>$ with $x \in A_{j'}$. But $j \in P_i \subseteq \sigma(i) = \{j'\}$, so $j = j'$ and $x \in A_j$. Assigning to each $j \in J$ the unique $x \in A_j$ obtained in this way yields a choice function on $\mathcal{A}$.

Reciprocally, if $f: J \longrightarrow \bigcup_{j \in J} A_j$ is a choice function on $\mathcal{A}$, define P_i for $i \in I$ by $P_i = \{j\}$ if $x = f(j)$, $P_i = \varnothing$ if $x \neq f(j)$, where $i = <j, x>$. Clearly $\{P_i: i \in I\}$ satisfies the required conditions.

The moral is: while the assertion *every nonempty set has an element* is (in classical logic) a truism, to assert it in the context of Boolean fuzzy sets is equivalent to asserting **AC**.

- **Every divisible abelian group is injective**[1]**.** An Abelian group G is *injective* if, for any Abelian group B, and any subgroup A of B, any homomorphism $A \to G$ can be extended to a homomorphism $B \to G$. G is *divisible* if, for any $a \in G$, and any integer $n \neq 0$, there is an element $b \in G$ for which $a = nb$. Given a

[1] Proved equivalent to **AC** by Blass [1979].

divisible group G, a subgroup A of an Abelian group B, and a morphism $h: A \to G$, consider the set $\mathcal{F}$ of pairs $\langle D, f\rangle$, where D is a subgroup of B containing A, and $f: D \to G$ is a morphism extending h. Then $\mathcal{F}$, partially ordered by extension, is (strongly) inductive, and so by **ZL** has a maximal element $\langle M, g\rangle$. We show that $M = B$, from which the injectivity of G immediately follows.

Suppose that $M \neq B$. Choosing an element $a \in B \setminus M$, let M_a be the subgroup of B generated by $M \cup \{a\}$: each element u of M_a is then of the form $x_u + n_u a$, with $x_u \in M_a$ and $n_u \in \mathbf{Z}$. There are then two cases: (i) $na \notin M$ for all $n \neq 0$, and (ii) $na \in M$ for some $n \neq 0$. In case (i), let $g': M_a \to G$ be defined by $g'(u) = g(x_u)$. Then the pair $\langle M_a, g'\rangle$ is a member of F properly extending $\langle M, g\rangle$, contradicting the maximality of the latter. In case (ii), there is a least positive integer n_0 for which $n_0 a \in M$. Then $g(n_0 a) \in G$ and so there is $a^* \in G$ for which $n_0 a^* = g(n_0 a)$. If we now define $g': M_a \to G$ by $g'(u) = g(x_u) + n_u a^*$, then the pair $\langle M_a, g'\rangle$ is a member of $\mathcal{F}$ properly extending $\langle M, g\rangle$, again contradicting the maximality of the latter. Accordingly $M = B$ and the result is proved.

- **In a commutative ring with identity, any (proper) ideal can be extended to a maximal (and hence prime) ideal.**[1] This follows quickly, using **ZL**, from the easily established fact that the family of ideals in such a ring is closed under unions of chains.
- **Every field has an algebraic closure.**[2] Recall that a field F is *algebraically closed* if every nonconstant polynomial in $F[x]$ has a zero in F, and that an *algebraic closure* of a field K is an

[1] Proved equivalent to **AC** by Hodges [1979].

[2] (Steinitz [1910]). The simple proof given here is based on that formulated by Jelonek [1993]. The assertion is also a consequence of the compactness theorem for first-order logic, and hence of **BPI**, which is weaker than **AC**.

algebraically closed field which is an algebraic extension of K. To show that each field has an algebraic closure, we shall need the elementary algebraic fact that, for any field K, and any nonconstant polynomial $f \in K[x]$, there is an algebraic extension L of K in which f has a zero. In fact, one need only take L to be the quotient field $K[x]/(f)$.

For each pair (K, f) consisting of a field K and a polynomial f over K, write $(K, f)^*$ for $K[x]/(f)$.

A set-theoretic argument using the Axiom of Replacement shows that there exists a set S such that $K \in S$ and for any field $L \in S$ and any polynomial $f \in L[x]$, $(L, f)^* \in S$. Now let $\mathscr{F} = \{L \in S$: *L is an algebraic extension of K*$\}$. Partially order $\mathscr{F}$ by stipulating that $L \sqsubseteq L'$ if L' is an algebraic extension of L. It is readily shown that $\mathscr{F}$ is closed under unions of chains and so **ZL** yields a maximal element M in $\mathscr{F}$. We shall show that M is an algebraic closure of K. Since M is an algebraic extension of K, it is enough to show that M is algebraically closed. Let $f \in M[x]$; then f has a zero, a say, in $(M, f)^*$. But the latter is a member of S, and hence also of $\mathscr{F}$, which extends M. Since M is maximal, $(M, f)^* \subseteq M$ and so *a fortiori* $a \in M$. So f has a zero in M and the latter is algebraically closed.

- **Every extension field has a transcendence basis.** A *transcendence basis* for an field G over a subfield F is a subset S of G which is algebraically independent over F and is also such that G is algebraic over the subfield $F(S)$ generated by $F \cup S$. To show that G has a transcendence basis over F, one observes that the family of independent sets is closed under unions of chains, then applies **ZL** to yield a maximal independent set, and finally notes that any maximal independent set is a transcendence basis. A similar argument yields the stronger result that any algebraically independent set can be extended to a transcendence basis.

- **Any automorphism of a subfield of an algebraically closed field A can be extended to the whole of A.** In proving this **ZL** is actually applied twice. We shall require three facts. Let F and G be two subfields of A, and let φ be an isomorphism between F and G. For each polynomial $p(x)$ over F let p^φ —the *φ-transform* of p—be the polynomial over G obtained by applying φ to the coefficients of p.

Fact I. If $a \in A$ is algebraic over F with minimal polynomial p, then for any zero b of p^φ there is an isomorphism extending φ of the subfields $F(a)$ and $G(b)$ generated by $F \cup \{a\}$ and $G \cup \{b\}$ respectively.

Fact II. If $a, b \in A$ are transcendental over F, G respectively, then there is an isomorphism of $F(a)$ and $G(b)$ extending φ and sending a to b.

In both cases the isomorphism in question is given by

$$p(a)/q(a) \mapsto p^\varphi(b)/q^\varphi(b).$$

Fact III. For each subfield F of A, let F^* consist of all the elements of A which are algebraic over F. Then F^* is an algebraically closed subfield of A.

Using these facts, we first show that

(*) *any isomorphism φ between subfields F and G of A can be extended to an isomorphism between F^* and G^*.*

To prove this, we apply **ZL** to the set $\mathscr{F} = \{\theta$: θ *is an isomorphism extending* φ *between a subfield of* F^* *and a subfield of* $G^*\}$. Partially ordered by inclusion, $\mathscr{F}$ is closed under unions of chains and so **ZL** yields a maximal element μ of $\mathscr{F}$. We show that domain(μ) = F^* and range(μ) = G^*.

If domain(μ) $\neq F^*$, choose a in $F \setminus$ domain(μ). Since a is algebraic over F and (according to Fact III) G^* is algebraicallyclosed, there is at least one $b \in G^*$ which is a zero of the μ-transform of the minimal polynomial of a over F. Thus by Fact I there is at least one way of extending μ to a larger isomorphism still in $\mathscr{F}$. This contradicts the maximality of μ and shows that domain(μ) = F^*.

Now since F^* is algebraically closed and μ is an isomorphism, range(μ) is an algebraically closed subfield of G containing G. But the only such subfield of G^* is G itself; hence range(μ) = G^*, and we are done.

Finally we can show, again using **ZL**, that any automorphism φ of a subfield of A can be extended to an automorphism of A. To this end let $\mathscr{F} = \{\theta$: *θ is an automorphism extending φ to some subfield of A*$\}$. Ordered by inclusion, $\mathscr{F}$ is closed under unions of chains and so by **ZL** has a maximal element μ. We must show that F = domain(μ) = A. If not, choose $a \in A \setminus F$. If a is algebraic over F, then $F^* \neq F$ and by (*) above μ can be extended to an automorphism of F^*, contradicting the maximality of μ. If a is transcendental over F, then, by Fact II, μ can be extended to an automorphism of $F(a)$. This again contradicts the maximality of μ. So there can be no element of A outside F and the proof is complete.

Remark. The *fundamental theorem of algebra* (whose proof does not require **AC** in any form) asserts that the field $\mathbb{C}$ of complex numbers is algebraically closed. Therefore **ZL** implies that any automorphism of a subfield of $\mathbb{C}$ is extensible to an automorphism of $\mathbb{C}$[1]. Thus, for example, the automorphism of $\mathbb{Q}(\sqrt{5})$ which sends $\sqrt{5}$ to $-\sqrt{5}$ can be extended to an automorphism of $\mathbb{C}$ with the same property. Also, since *e* and π are transcendental over the rational field $\mathbb{Q}$ we may take transcendence bases (whose existence is ensured by **ZL**) *S* and *T* of $\mathbb{C}$ over $\mathbb{Q}$ containing *e* and π respectively. All transcendence bases of $\mathbb{C}$ over $\mathbb{Q}$ have the same cardinality (that of the continuum), so there is a bijection between *S* and *T* which sends *e* to π. The algebraic independence of *S* and *T* enables this bijection to be extended to an isomorphism between $\mathbb{Q}(S)$ and $\mathbb{Q}(T)$ and, using (*) above, this isomorphism in turn extends to an isomorphism θ of $\mathbb{Q}(S)^*$ and $\mathbb{Q}(T)^*$. Since $\mathbb{Q}(S)^* = \mathbb{Q}(T)^* = \mathbb{C}$, θ is an automorphism of $\mathbb{C}$ sending *e* to π.

In fact any permutation of a transcendence basis of $\mathbb{C}$ over $\mathbb{Q}$ extends to an automorphism of $\mathbb{C}$. Since any such transcendence basis has cardinality $2^{\aleph_0}$, there are $(2^{\aleph_0})^{2^{\aleph_0}} = 2^{2^{\aleph_0}}$ such permutations, and hence also $2^{2^{\aleph_0}}$ automorphisms[2] of $\mathbb{C}$. This is one of the most remarkable consequences of **ZL** (or **AC**). For consider the fact that, in the absence of **AC**, one can exhibit

[1] The problem of the existence of nontrivial automorphisms of $\mathbb{C}$ was propounded by C. Segre in 1889 in connection with the question of the existence of non-projective collineations in a bicomplex plane.

[2] I learned recently that the group of $2^{2^{\aleph_0}}$ automorphisms of $\mathbb{C}$ is known as the *absolute Galois group*.

just *two* automorphisms of ℂ, namely the identity and conjugation ($x + iy \mapsto x - iy$) . Now it is known that the presence of a *single* automorphism of ℂ different from either of these entails the existence of Lebesgue nonmeasurable subsets of the continuum[1]. So it follows from Solovay's construction[2] of a model of set theory in which all subsets of the continuum are Lebesgue measurable that without **AC** *none* of these $2^{2^{\aleph_0}}$ automorphisms of ℂ —apart from identity and conjugation— necessarily exist. Yet what might be called the concrete traces of these "fugitive" automorphisms of ℂ are often identifiable, as can be seen from the example above of the $\mathbb{Q}(\sqrt{5})$ automorphism. While the action on $\mathbb{Q}(\sqrt{5})$ of any extension φ to ℂ (whose existence is guaranteed by **AC)** is perfectly clear, its action on the rest of ℂ is decidedly otherwise - and this despite the fact that since the end of the 18th century ℂ itself has been regarded as a perfectly definite mathematical object. Indeed, aside from some general facts concerning nontrivial automorphisms of ℂ (for example that they must send at least some real numbers to complex ones, that they are everywhere discontinuous, andthat they map discs to nonmeasurable sets in the complex plane) all one knows about φ is that it extends the $\mathbb{Q}(\sqrt{5})$ automorphism.

---

[1] See, e.g., Kestelman [1951]
[2] Solovay [1970].

- **Every real field has a real closure.** A field K is said to be *real* if –1 is not a sum of squares in K. A field is said to be *real closed* if it is real, and if any algebraic extension of K which is real must coincide with K. In other words, a real closed field is maximal with respect to the property of being a real subfield of an algebraic closure[1]. A *real closure* of a real field K is a real closed field which is algebraic over K.

Now to show that every real field has a real closure, let A be an algebraic closure of K and consider the set $\mathscr{F}$ of subfields F of A which are both real and extend K. Then $\mathscr{F}$, partially ordered by inclusion, is closed under unions of chains and so by **ZL** has a maximal element. This latter is real closed and, as a subfield of A, algebraic over K.

- **Tychonov's Theorem[2]—the product of compact topological spaces is compact.** Let $\{X_i: i \in I\}$ be a family of compact spaces. To show that their product $\prod_{i \in I} X_i = X$ is compact it suffices to show that, if $\mathscr{F}$ is any family of closed subsets of X with the *finite intersection property (fip)*—that is, satisfying the condition that the intersection of any finite subfamily is nonempty—, then $\bigcap \mathscr{F} \neq \varnothing$. So let $\mathscr{F}$ be such a family and let $\mathfrak{Z}$ be the collection of all families of subsets of X which include $\mathscr{F}$ and have the *fip*. It is a simple matter to verify that $\mathfrak{Z}$ is closed under unions of chains, and so by **ZL** it has a maximal member $\mathscr{M}$. We show that $\bigcap_{M \in \mathscr{M}} \overline{M} \neq \varnothing$.

First of all observe that $\mathscr{M}$ satisfies the two following conditions:

---

[1] This definition comes from Lang [2002].

[2] Tychonov [1935].

(i) $M_1, \ldots, M_n \in \mathcal{M} \Rightarrow M_1 \cap \ldots \cap M_n \in \mathcal{M}$;

(ii) $A \subseteq X$ & $A \cap M \neq \varnothing$ for all $M \in \mathcal{M} \Rightarrow A \in \mathcal{M}$.

To verify (i): if $M_1, \ldots, M_n \in \mathcal{M}$, then clearly the family $\mathcal{M} \cup \{M_1 \cap \ldots \cap M_n\}$ is a member of $\mathfrak{Z}$; since it includes $\mathcal{M}$, and $\mathcal{M}$ is maximal in $\mathfrak{Z}$, it must coincide with $\mathcal{M}$, so that, *a fortiori*, $M_1 \cap \ldots \cap M_n$ must be a member of $\mathcal{M}$. For (ii), suppose that A is a subset of X which meets every member of $\mathcal{M}$. Then, for each finite subset $\{M_1, \ldots, M_n\}$ of $\mathcal{M}$ we have by (i) $M_1 \cap \ldots \cap M_n \in \mathcal{M}$, so that $A \cap (M_1 \cap \ldots \cap M_n) \neq \varnothing$. Therefore $\mathcal{M} \cup \{A\}$ has the *fip*,and so is a member of $\mathfrak{Z}$ including $\mathcal{M}$. The latter's maximality implies then that $A \in \mathcal{M}$. This proves (ii).

Now write π_i for the (continuous) projection of $\prod_{i \in I} X_i$ onto X_i. Then for each $i \in I$ the family $\{\overline{\pi_i[M]} : M \in \mathcal{M}\}$ of closed subsets of the compact space X_i has the *fip* (since $\mathcal{M}$ itself does) and hence nonempty intersection. For each $i \in I$ choose a member x_i of this intersection. Then $x = (x_i: i \in I) \in X$ has the property that each open neighbourhood U of x_i meets $\pi_i[M]$, and so $\pi_i^{-1}[U]$ meets M, for any $M \in \mathcal{M}$. Therefore, by (ii), $\pi^{-1}[U] \in \mathcal{M}$. It follows now from (i) that, for any open neighbourhoods $U_1, \ldots, U_n$ of $x_{i_1}, \ldots, x_{i_n}$ respectively, $\pi_{i_1}^{-1}[U_1] \cap \ldots \cap \pi_{i_n}^{-1}[U_n] \in \mathcal{M}$. In other words, every basic neighbourhood of x is a member of $\mathcal{M}$. Since $\mathcal{M}$ has the *fip*, each basic neighbourhood of x meets each member of $\mathcal{M}$, that is, x is in the closure of each member of $\mathcal{M}$. Thus $x \in \bigcap_{M \in \mathcal{M}} \overline{M} \neq \varnothing$.

Finally, since each member of $\mathcal{F}$ is closed, and $\mathcal{F} \subseteq \mathcal{M}$, it follows that $\bigcap_{M \in \mathcal{M}} \overline{M} \subseteq \bigcap \mathcal{F}$, so that $\bigcap \mathcal{F} \neq \varnothing$ and the result follows.

Tychonov's theorem is actually equivalent to **AC.** It is interesting to note that, like the derivation of **AC** from **ZL,** the derivation of **AC** from Tychonov's theorem is remarkably straightforward. Here is surely the simplest derivation[1]. Given an indexed family of nonempty sets $\{X_i: i \in I\}$, let a be an element such that $a \notin \bigcup_{i \in I} X_i$, and for each $i \in I$ let $Y_i = X_i \cup \{a\}$. Topologize each Y_i by declaring just the subsets $\varnothing$, $\{a\}$, Y_i to be open. Evidently each space Y_i is then compact and so, by Tychonov's theorem, the product $\prod_{i \in I} Y_i$ is also. Since each X_i is closed in Y_i, $\pi_i^{-1}[X_i]$ is closed in $\prod_{i \in I} Y_i$; and it is easily shown that the family $\{\pi_i^{-1}[X_i] : i \in I\}$ has the *fip*. Hence its intersection, which clearly coincides with $\prod_{i \in I} X_i$, is nonempty.

The original derivation of **AC** from Tychonov's theorem, due to Kelley [1950], used more complicated topologies, but each was T_1 ("points are closed"), so showing that **AC** is derivable from Tychonov's theorem restricted to spaces satisfying this natural condition. Using the above notation, in Kelley's derivation each Y_i is topologized by first equipping X_i with the so-called *cofinite* topology, that is, by declaring open, along with $\varnothing$ and X_i, all complements of finite ("cofinite") subsets thereof, and then regarding Y_i as a *one-point compactification* of X_i. This amounts to assigning to each Y_i the topology consisting of the subsets $\varnothing$, $\{a\}$, all cofinite subsets of X_i, the unions of these with $\{a\}$, and Y_i. Each resulting space is then both T_1 and compact, and the argument goes through as above.

[1] Alas [1969].

But for compact *Hausdorff* spaces Tychonov's theorem is equivalent to **BPI**[1] and hence is weaker than **AC.**

- **Every lattice with a largest element has a maximal (proper) ideal (or, equivalently every lattice with a least element has a maximal filter).** Just as for rings, this assertion follows quickly from **ZL**. It was proved equivalent to **AC** by Scott [1954]. Later the corresponding assertion for distributive lattices was proved by Klimovsky [1958], and for lattices of sets by Bell and Fremlin [1972]. The best result along these lines so far is due to Herrlich [2002], who shows that **AC** holds iff

(#) *the lattice of closed subsets of any nonempty topological space contains a maximal (proper) filter.*

AC2 can be derived from **(#)** as follows. For each topological space X write $\mathscr{C}X$ for the lattice of closed subsets of X. Now let $\mathfrak{A} = \{A_i: i \in I\}$ be an indexed family of nonempty sets. Choose an individual $*$ not contained in any A_i, and let $A_i^* = A_i \cup \{*\}$. Topologize each A_i^* by declaring A_i^* itself and any finite subset of A_i to be a closed subset. Clearly the product space $A^* = \prod_{i \in I} A_i^*$ is then nonempty, and it is easy to show that the minimal members of $\mathscr{C}A^*$ are precisely the singletons $\{a\}$ with a a choice function on $\mathfrak{A}$.

We now show that each maximal filter in $\mathscr{C}A^*$ is generated by a minimal closed set, and hence each determines a choice function on $\mathfrak{A}$. Write π_i for the (continuous) projection of $\prod_{i \in I} A_i^*$ onto A_i^*, let $\mathscr{F}$ be a maximal, hence prime, filter in $\mathscr{C}A^*$, and let $\mathscr{F}_i$ be the family of closed subsets X of A_i^* for which $\pi_i^{-1}[X] \in \mathscr{F}$. Each $\mathscr{F}_i$ is then a prime filter in $\mathscr{C}A_i^*$. Since A_i^* is obviously

[1] Rubin and Scott [1954].

compact, and $\mathscr{F}_i$ has the finite intersection property it follows that $C_i = \bigcap \mathscr{F}_i \in \mathscr{F}_i$ (in fact $\mathscr{F}_i$ is generated by C_i). Now let $J = \{i \in I : C_i \subseteq A_i\}$; it is easy to see that $i \in J$ iff there exists some finite subset X of A_i with $\pi_i^{-1}[X] \in \mathscr{F}$. If $i \in J$, then C_i is finite and so since $\mathscr{F}_i$ is prime there is a unique $a_i \in A_i$ such that $C_i = \pi_i^{-1}(a_i)$. If $i \notin J$, then $* \in C_i$.

Now define, for $i \in I$,

$$Z_i = \{a_i\},\ z_i = a_i \text{ if } i \in J$$
$$Z_i = A_i^*,\ z = * \text{ if } i \notin J,$$

and $Z = \prod_{i \in I} Z_i$. Then Z is the closure of $\{z\}$ in A^*.

We claim that every neighbourhood of z meets every member of $\mathscr{F}$. From this it will follow that $z \in \bigcap \mathscr{F}$, and hence $\varnothing \neq Z \subseteq \bigcap \mathscr{F}$. Since Z meets every member of $\mathscr{F}$, and the latter is maximal, $Z \in \mathscr{F}$, so that Z generates $\mathscr{F}$. From the maximality of $\mathscr{F}$ it follows that Z is minimal, and so determines a choice function on $\mathfrak{A}$.

Finally, to prove the claim, take $F \in \mathscr{F}$, $i \in I$ and let U be a neighbourhood of z_i in A_i^*. If $i \in J$, then $z_i = a_i$ and so $\pi_i^{-1}(z_i) = \pi_i^{-1}(a_i) \in \mathscr{F}$, whence $\pi_i^{-1}(z_i) \cap F \neq \varnothing$ so *a fortiori* $\pi_i^{-1}[U] \cap F \neq \varnothing$. On the other hand, if $i \notin J$, and $\pi_i^{-1}[U] \cap F = \varnothing$, then $F \subseteq \pi_i^{-1}[A_i^* \backslash U]$, so that $\pi_i^{-1}[A_i^* \backslash U] \in \mathscr{F}$. Since $A_i^* \backslash U$ is a finite subset of A_i^*, this violates the condition that $i \notin J$. Thus $\pi_i^{-1}[U]$ meets every member of $\mathscr{F}$; the claim now follows easily from the primeness of $\mathscr{F}$.

- **Stone-Čech compactification theorem: for each completely regular space X there is a compact Hausdorff space βX into which X can be densely embedded.**[1] Here βX is the space whose underlying set is the set of maximal ideals in the ring $C(X)$ of continuous real-valued functions of X (**ZL** is required to show that there are enough of these). The topology on βX – the *Stone-Zariski topology* – is defined by taking the family of sets $\{S(a): a \in X\}$ as a base, where $S(a) = \{M \in \beta X: a \notin M\}$. X is densely embedded in βX by the map $a \mapsto \{f \in C(X): f(a) = 0\}$.
- **Gelfand-Kolmogorov theorem: if X and Y are compact Hausdorff spaces and $C(X) \cong C(Y)$, then X is homeomorphic to Y.**[2] For, writing $\approx$ for "is homeomorphic to", if X and Y are compact Hausdorff and $C(X) \cong C(Y)$, then $X \approx \beta X \approx \beta Y \approx Y$.
- **Gelfand-Naimark-Stone theorem: each real C*-algebra is isomorphic to $C(X)$ for some compact Hausdorff space X.**[3] Here, given a C*-algebra A, the space X is the space of maximal ideals in A with the Stone-Zariski topology.
- **Every linear space has a basis**[4]**.** Here it is only necessary to observe that a basis for a linear space is precisely an inclusion-maximal independent subset, and that the family of all such subsets is closed under unions of chains, so that **ZL** yields a maximal member.
- **All bases of a linear space have the same cardinality.** Let B and C be bases of a linear space L. Without loss of generality it may be assumed that B and C are disjoint. By the symmetry of these assumptions, together with the Schröder-Bernstein theorem,

---

[1] Čech [1937], Stone [1937].

[2] Gelfand and Kolmogorov [1939]

[3] Gelfand [1939, 1941], Gelfand and Naimark [1943], Stone [1940].

[4] The essential idea behind the proposition is due to Hamel [1905]. It was proved equivalent to **AC** by Blass [1984].

it suffices to show that there is an injection of B into C. Let $\mathscr{F}$ be the set of pairs $<X, f>$ where $X \subseteq B$ and f is an injection $X \to C$ such that range(f) $\cup$ ($B \setminus X$) is a linearly independent set. Partially order $\mathscr{F}$ by extension. It is straightforward to show that $\mathscr{F}$ is then (strongly) inductive and hence by **ZL** has a maximal element $<M, g>$. We claim that $M = B$.

For suppose not. Then R = range(g) $\neq C$, for each element of $B \setminus M$ is linearly dependent on the basis C but not on R. That being the case, we may choose $c_0 \in C \setminus R$; then either c_0 is linearly independent of $R \cup (B \setminus M)$ or is dependent on it. In the former case, for arbitrary $b \in B \setminus M$, the pair $<M \cup \{b\}, g \cup \{<b,c_0>\}>$ is a member of $\mathscr{F}$ properly extending $<M, g>$, contradicting its maximality. In the latter case, c_0 can be represented as a finite sum

$$c_0 = \sum_{c \in R} \lambda_c c + \sum_{b \in M} \mu_b b,$$

where the λ_c and the μ_b are elements of the underlyincoefficient field. Because c_0 is independent of R, there must be at least one b, b_0 say, in this representation for which $\mu_{b_0} \neq 0$. Let g' be the map $g \cup \{<b_0, c_0>\}$. Then the pair $<M \cup \{b_0\}, g'>$ is a proper extension of $<M, g>$ which is also a member of $\mathscr{F}$, since the choice of b_0 ensures that range(g') $\cup$ ($B \setminus (M \cup \{b_0\})$) is linearly independent. This again contradicts the maximality of $<M, g>$.

We conclude that $M = B$, so that g is an injection of B into C, and the proof is complete.

- **The Hahn-Banach Theorem**[1]. Suppose that the real-valued function p on the linear space[2] L satisfies

[1] Originally proved in 1929, this theorem was later shown to be a consequence of **BPI** and hence weaker than **AC**.

[2] Henceforth all linear spaces will be presumed to have the real numbers as scalar field.

$$p(x + y) \leqslant p(x) + p(y), p(\alpha x) = \alpha p(x) \quad \text{for } \alpha \geqslant 0, x, y \in L.$$

Let f be a linear functional defined on a subspace K of L such that $f(x) \leqslant p(x)$ for $x \in K$. Then there is a linear functional F on L extending f such that $F(x) \leqslant p(x)$ for $x \in L$.

To prove this from **ZL**, let $\mathscr{F}$ be the set of all pairs $\langle X, g \rangle$ consisting of a subspace X of L containing K and a linear functional g on X extending f for which the inequality $g(x) \leqslant p(x)$ holds for all $x \in X$, Then $\mathscr{F}$, partially ordered by extension, is (strongly) inductive and so **ZL** applies to yield a maximal member $\langle M, F \rangle$. Thus F is a linear extension of f such that $F(x) \leqslant p(x)$ for all $x \in M$. It remains to show that $M = L$.

For contradiction's sake, suppose that there is a point u in L which is not in M. Then any point in the subspace U of L generated by $M \cup \{u\}$ has a unique representation in the form $z + \alpha u$. For any constant γ, the function G defined on U by setting

$$G(z + \alpha u) = F(z) + \alpha\gamma$$

is a linear functional properly extending F. The desired contradiction will be obtained and the proof completed if we can show that γ can be chosen in such a way that

(*) $$G(x) \leqslant p(x) \quad \text{for all } x \in U.$$

Let $x, y \in U$; then the inequality

$$F(y) - F(x) = F(y - x) \leqslant p(y - x) \leq p(y + u) + p(-u - x)$$

gives

$$-p(-u - x) - F(x) \leqslant p(y + u) - F(y).$$

Since the left-hand side of this last inequality is independent of y and the right hand side is independent of x, there is a constant γ such that

$$\text{(i) } \gamma \leqslant p(y+u) - F(y) \quad \text{(ii) } -p(-u-y) - F(y) \leqslant \gamma,$$

for $y \in Z$. For $x = z + \alpha u$ in U, the inequality

$$G(x) = F(z) + \alpha\gamma \leqslant p(z) + \alpha u = p(x),$$

which holds for $\alpha = 0$ by hypothesis, is obtained for $\alpha > 0$ by replacing y by $\alpha^{-1}z$ in (i), and for $\alpha < 0$ by replacing y by $\alpha^{-1}z$ in (ii).

Thus we obtain (*) in all cases, and hence the required contradiction.

The Hahn-Banach theorem has numerous consequences. We shall require one in particular for linear topological spaces. A subset A of a linear space L is *convex* if, for arbitrary $x, y \in A$, $\alpha x + (1 - \alpha)y \in A$ whenever $0 \leqslant \alpha \leqslant 1$. Now suppose that L is a topological linear space. L is said to be *locally convex* if 0 has a neighbourhood base consisting of open convex sets. (Note that every normed space is locally convex with the norm topology.) Then the **Separation Principle for Locally Convex Spaces**, which can be proved from the Hahn-Banach theorem, asserts that, if L is a locally convex Hausdorff linear topological spaces, then, for any distinct points x and y of L, there is a continuous linear functional f on L such that $f(x) \neq f(y)$.

- **The Krein-Milman Theorem—a compact, convex subset of a locally convex Hausdorff linear topological space has at least one extreme point.** Let us call an *extreme subset* of a convex subset A of a linear space any closed subset $X \subseteq A$ such that, for any $x, y \in A$, if $\alpha x + (1 - \alpha)y \in X$ for some $0 < \alpha < 1$, then both x and y belong to X. An *extreme point* of A is an element e of A for which $\{e\}$ is an extreme subset of A. It is easily verified that e is an extreme point of A if and only if it belongs to no open line segment in A, that is, e can be represented in the form $\alpha x + (1 - \alpha)y$ with $x, y \in A$ with $0 \leqslant \alpha \leqslant 1$ only when $\alpha = 0$ or $\alpha = 1$.

Before deriving the Krein-Milman theorem we note the following fact: if A is a nonempty compact convex subset of a linear topological space L, and f is a continuous linear functional on L, then, writing β for $\inf f[A]$, the set $B = A \cap f^{1}(\beta)$ is a nonempty extreme subset of A. For the continuity of f ensures both that B is closed, and that f attains its infimum on A, so that $B \neq \varnothing$. Finally, suppose that $x, y \in B$ and $(1 - \alpha)x + \alpha y \in B$ with $0 < \alpha < 1$. Then both x and y belong to B. For if $x \notin B$, then $f(x) > \beta$ so that

$$f((1 - \alpha)x + \alpha y) = (1 - \alpha)f(x) + \alpha f(y) > (1 - \alpha)\beta + \beta = \beta,$$

which contradicts the hypothesis that $(1 - \alpha)x + \alpha y \in B$. Thus x must belong to B. Similarly $y \in B$. Accordingly B is an extreme subset of B.

Now let L be a locally convex Hausdorff linear topological space, and A a closed convex subset of L. Let $\mathscr{E}$ be the set of all nonempty extreme subsets of A, partially ordered by inclusion. By the above fact, $\mathscr{E}$ is nonempty. Also $\mathscr{E}$ is reductive, since if $\mathscr{C}$ is any chain in $\mathscr{E}$, $\bigcap\mathscr{C}$ is extreme, nonempty since A is compact, and hence a lower bound for $\mathscr{C}$ in $\mathscr{E}$. So by **DZL** $\mathscr{E}$ has a minimal member E. We claim that E is a singleton. For otherwise E would contain two distinct points x and y. By the Separation Principle, there is a continuous linear functional on L such that $f(x) < f(y)$. By the fact above, $B = E \cap f^{1}(\inf f[E])$ is a nonempty extreme subset of E which does not contain y, contradicting the minimality of E. So E is a singleton, and its solitary element is an extreme point of A.

It has been shown[1] that, while **AC** cannot be derived from the Krein-Milman theorem alone, it can be derived from the assertion that the unit ball of the dual of a real normed linear space has an extreme point, which is itself a consequence of the Krein-Milman theorem + **BPI.** There it is shown that, given any indexed family $\mathfrak{A} = \{A_i\colon i \in I\}$ of nonempty sets, there is a correspondence between choice functions on $\mathfrak{A}$ and the extreme points of the unit ball of a certain real normed linear space $L(\mathfrak{A})$ (itself the dual of another normed linear space) constructed from $\mathfrak{A}$. Writing A for $\bigcup_{i\in I} A_i$, $L(\mathfrak{A})$ is the linear space

$$\{x \in \mathbb{R}^A : \sup_{i\in I} \sum_{t\in A_i} |\, x(t)\,| < \infty\}$$

with the norm $\|x\| = \sup_{i\in I} \sum_{t\in A_i} |\, x(t)\,|$.

Let $B(\mathfrak{A}) = \{x\in L(\mathfrak{A})\colon \|x\| \leqslant 1\}$ be the unit ball of $L(\mathfrak{A})$. We describe a natural bijection between extreme points of $B(\mathfrak{A})$ and the set $\mathbf{P}I \times \prod_{i\in I} A_i$ [2].

Given a subset $J \subseteq I$, and a choice function f on $\mathfrak{A}$, the extreme point e_J correlated with $<J, f>$ is obtained by setting $e_J(f(i)) = 1$ for $i \in J$, $e_J(f(i)) = -1$ for $i \in I \setminus J$, and $e_J(t) = 0$ for $t \in A \setminus \{f(i) : i \in I\}$.

Inversely, let e be an extreme point of $B(\mathfrak{A})$. We are going to show that, for each $i \in I$, there is a unique $t^* \in A_i$ such that $|\, e(t^*)\,| = 1$ and $e(t) = 0$ for all $t \in A_i \setminus \{t^*\}$.

---

[1] Bell and Fremlin [1972].

[2] Here $\mathbf{P}I$ is the power set of I; also recall that $\prod_{i\in I} A_i$ is the set of choice functions on $\{A_i\colon i \in I\}$.

We show first that, for any $i \in I$, e cannot take the value 0 everywhere on A_i. For if it did, choose a point $t_0 \in A_i$ and define $x, y \in B(\mathfrak{A})$ by $x(t) = y(t) = e(t)$ for $t \in A - \{t_0\}$, $x(t_0) = 1$, $y(t_0) = -1$. Then $x \neq e \neq y$ and $e = \frac{1}{2}(x+y)$, contradicting the extremeness of e. Thus e does not take the value 0 everywhere on A_i.

Next, we show that e is nonzero at exactly one point in A_i. For suppose that $e(t_0) \neq 0 \neq e(t_1)$ for two distinct points t_0, $t_1 \in A_i$. Define $x, y \in B(\mathfrak{A})$ by

$$x(t) = y(t) = e(t) \text{ for } t \in A \setminus \{t_0, t_1\},$$
$$x(t_0) = e(t_0)(1 + |e(t_1)|) \qquad x(t_1) = e(t_1)(1 - |e(t_0)|)$$
$$y(t_0) = e(t_0)(1 - |e(t_1)|) \qquad y(t_1) = e(t_1)(1 + |e(t_0)|).$$

Then $x \neq e \neq y$ and $e = \frac{1}{2}(x + y)$, again contradicting the extremeness of e.

Thus there is a unique $t^* \in A_i$ for which $e(t^*) \neq 0$ and $e(t) = 0$ for all $t \in A_i \setminus \{t^*\}$. And in fact $|e(t^*)| = 1$. For if $|e(t^*)| < 1$, define $x, y \in B(\mathfrak{A})$ by $x(t) = y(t) = e(t)$ for $t \in A \setminus \{t^*\}$, $y(t^*) = 0$ and $x(t^*) = +1$ or -1 according as $e(t^*) > 0$ or $e(t^*) < 0$. Then, writing $\alpha = |e(t^*)|$, we have $0 < \alpha < 1$ and $e = \alpha x + (1 - \alpha)y$, yet again contradicting the extremeness of e.

Accordingly for each $i \in I$, there is a unique $t^* \in A_i$ such that $|e(t^*)| = 1$ and $e(t) = 0$ for all $t \in A_i \setminus \{t^*\}$. Let f be the choice function on $\mathfrak{A}$ defined by setting $f(i)$ to be this unique $t^* \in A_i$; and let $J = \{i \in I: e(t^*) = 1\}$. Finally, we correlate the pair $\langle J, f\rangle \in \mathbf{P}I \times \prod_{i \in I} A_i$ with e.

It should be clear that the foregoing procedure establishes the required bijection.

- **Model Existence Theorem for first-order logic**[1]**: each consistent first-order theory has a model.** This was shown by

[1] Gödel [1930]. 17], Henkin [1954].

Henkin [1954] to be equivalent to **BPI**, and hence weaker than **AC.** If the cardinality of the model is specified in the appropriate way, the assertion becomes equivalent to **AC.** The model existence theorem is proved by first using **ZL** to produce a maximal consistent extension *M* of a given consistent first-order theory *T* and then noting that *M* is complete, that is, any sentence of the language of *M* is provable or refutable from *M*. A model of *T* is then constructed from *M*.[1]

- **Compactness Theorem for First-Order Logic[2]—if every finite subset of a of a set of first-order sentences has a model, then the set has a model[3].** While this is an immediate consequence of the Model Existence Theorem, the compactness theorem also admits a proof from Łoś's Theorem + **BPI** which is free of syntactic notions (such as consistency) and whose elegance recommends it for presentation.

Thus suppose that each finite subset Δ of a given set Σ of first-order sentences has a model $\mathfrak{A}_\Delta$; for simplicity write *I* for the family of all finite subsets of Σ. For each $\Delta \in I$ let $\Delta^* = \{\Phi \in I: \Delta \subseteq \Phi\}$. For any members $\Delta_1, \ldots, \Delta_n$ of *I*, we have $\Delta_1 \cup \ldots \cup \Delta_n \in \Delta_1{}^* \cap \ldots \cap \Delta_n{}^*$ and so the collection $\{\Delta^*: \Delta \in I\}$ has the finite intersection property. From **BPI** it follows that it can be extended to an ultrafilter *U* in **P***I*. The ultraproduct $\prod_{\Delta \in I} \mathfrak{A}_\Delta / U$ is then a model of Σ. For if $\sigma \in \Sigma$, then $\{\sigma\} \in \Delta$ and $\mathfrak{A}_{\{\sigma\}} \models \sigma$; moreover, $\mathfrak{A}_\Delta \models \sigma$ whenever $\sigma \in \Delta$. Hence $\{\sigma\}^* = \{\Delta \in I: \sigma \in \Delta\} \subseteq$

[1] For details see, e.g. Bell and Machover [1977].

[2] Gödel [1930], Malcev [1937], others.

[3] The compactness theorem was shown by Henkin in 1954 to be equivalent to **BPI**, and is accordingly weaker than **AC.**

$\{\Delta \in I: \mathfrak{A}_\Delta \models \sigma\}$. Since $\{\sigma\}^* \in U$, $\{\Delta \in I: \mathfrak{A}_\Delta \models \sigma\} \in U$ and therefore, by Łoś's Theorem, $\prod_{\Delta \in I} \mathfrak{A}_\Delta / \mathscr{U} \models \sigma$. The proof is complete.

SOME CONSTRUCTIVE EQUIVALENTS AND CONSEQUENCES OF ZL

We have seen that, in set theory based on classical logic, **ZL** is equivalent to **AC**. But in set theory based on intuitionistic logic, in which the Law of Excluded Middle is not assumed, the situation is decidedly otherwise. There, **ZL** turns out to be remarkably weak: not only does it fail to imply **AC**, but one cannot even prove from it, for example, the Boolean Prime Ideal theorem or the Stone Representation Theorem for Boolean Algebras. This is because, as we show in Chapter VI, **ZL** has no nonconstructive purely logical consequences, while both **AC** and the Stone Representation Theorem imply the Law of Excluded Middle, the Boolean prime ideal theorem implies the nonconstructive form of de Morgan's law: both of these latter facts are established in Chapter V. In fact, the vast majority of the assertions constructively provable from **ZL** make explicit mention of the notion of maximality: for example, the Hausdorff Maximal Principle, which we have noted is in fact constructively equivalent to it. So it is of interest to seek set-theoretical propositions which are constructively equivalent to, or at least constructively provable from, **ZL** but whose formulations do not make reference to maximality.

First, we note again that the proof of the Sikorski Extension Theorem for Boolean Algebras from **ZL** is constructively sound.

As another example, let us consider Tychonov's theorem in a familiar restricted form: namely, the product of compact *Hausdorff* spaces is compact. We shall see that, if the topological terms involved are provided with suitable constructive formulations, this form of Tychonov's theorem is a constructive consequence of **ZL.**

To begin with, we shall construe "nonempty" in the positive sense of being "inhabited": thus a set A is *inhabited* if $\exists x.\ x \in A$. The *closure* $\overline{A}$ of a subset A a topological space X is defined to be the set of all $x \in X$ such that, for any open neighbourhood U of x, $U \cap A$ is inhabited. A is *closed* if $A \subseteq \overline{A}$. X will be called *Hausdorff* if, for any $x, y \in X$, whenever $U \cap V$ is inhabited for every pair of open neighbourhoods U of x and V of y, then $x = y$. Recall that a family $\mathcal{F}$ of sets has the *finite intersection property* if the intersection of any finite subfamily of $\mathcal{F}$ is inhabited. Finally the topological space X is *compact* if, for any family $\mathcal{F}$ of closed subsets of X with the finite intersection property, $\bigcap \mathcal{F}$ is inhabited.

Now let us reexamine the demonstration of Tychonov's theorem from Chapter III. If in it we employ the above definitions of "Hausdorff" and "compact" and replace "$\neq \varnothing$" (i.e. "nonempty") by "inhabited", we find that what results is constructively sound (modulo the use of **ZL**) except for the single application of **AC** to select, for each $i \in I$, a member x_i of the (inhabited) intersection of the family $\{\overline{\pi_i[M]} : M \in \mathcal{M}\}$ of closed subsets of the compact space X_i. Now if each such intersection happens to be a *singleton*, then the use of **AC** becomes eliminable, and as a result the demonstration from **ZL** will be constructively sound. We show that this is the case when each X_i is Hausdorff. For assuming the latter, suppose that $x, y \in \bigcap\{\overline{\pi_i[M]} : M \in \mathcal{M}\}$. Then, for each pair of open neighbourhoods U of x, V of y, and any $M \in \mathcal{M}$, both $\pi_i[M] \cap U$ and $\pi_i[M] \cap V$ are inhabited, and so therefore are $M \cap \pi_i^{-1}[U]$ and $M \cap \pi_i^{-1}[V]$. It now follows from property (ii) of $\mathcal{M}$ that both $\pi_i^{-1}[U]$ and $\pi_i^{-1}[V]$ are members of $\mathcal{M}$, and so, since $\mathcal{M}$ has the finite intersection property, $\pi_i^{-1}[U] \cap \pi_i^{-1}$

$[V] = \pi_i^{-1}[U \cap V]$ must be inhabited. It follows that $U \cap V$ is inhabited. Since this is true for arbitrary open neighbourhoods U, V, and X_i is Hausdorff, we conclude that $x = y$. So $\bigcap\{\overline{\pi_i[M]} : M \in \mathcal{M}\}$ is a singleton, and we are done.

To sum up, *Tychonov's theorem for compact Hausdorff spaces is constructively derivable from* **ZL**.

We conclude this chapter with an account of some propositions not explicitly involving maximality which are constructively *equivalent* to **ZL**.[1] To formulate them we shall require a number of definitions.

Let $(P, \leqslant)$ be a poset. If a subset X of P has a greatest lower bound (respectively least upper bound) it will be written $\bigwedge X$ (respectively $\bigvee X$). P is *complete* if $\bigwedge X$ and $\bigvee X$ exist for every subset X. A subset B of P is a *base* for P if, for any $x, y \in P$, we have

$$\forall b \in B[b \leq x \Rightarrow b \leq y] \Rightarrow x \leq y.$$

Notice that if P is complete, B is a base iff

$$\forall x \in L.\ x = \bigvee \{b \in B : b \leq x\}.$$

A map $f: P \to P$ is (i) *self-adjoint* if for any $x, y \in P$ we have

$$x \leq f(y) \Leftrightarrow y \leq f(x),$$

and (ii) *inflationary* on a subset $X \subseteq P$ if $x \leq f(x)$ for all $x \in X$.

Lemma. *Let P be a poset and $f: P \to P$ a self-adjoint map. Let X be a subset of P for which $\bigvee X$ exists. Then $\bigwedge f[X]$ exists and in fact coincides with $f(\bigvee X)$.*

Proof. We have, for any $y \in P$

$$\forall x \in X.\ y \leq f(x) \Leftrightarrow \forall x \in X.\ x \leq f(y) \Leftrightarrow \bigvee X \leq f(y) \Leftrightarrow y \leq f(\bigvee X).$$

It follows in particular that any self-adjoint map on a poset is order-inverting.

---

[1] Bell [2003].

We use this to establish what we shall term the

Fixed Point Property for self-adjoint maps (FP). *Assume* **ZL.** *Let $f: P \to P$ be a self-adjoint map on a complete poset P possessing a base B on which f is inflationary. Then f has a fixed point.*

Proof. Let $D = \{x \in P: x \leq f(x)\}$. We claim that, with the order inherited from P, D is inductive. For consider any chain C in D, and let $c = \bigvee C$. We claim that $c \in D$. To prove this, we note that $f(c) = f(\bigvee C) = \bigwedge f[C]$ by the lemma above , so it suffices to show that $c \leq \bigwedge f[C]$, i.e. $x \leq f(y)$ for all $x, y \in C$. Now if $x, y \in C$, then either $x \leq y$ or $y \leq x$. In the first case $x \leq y \leq f(y)$; in the second $f(x) \leq f(y)$ so that $x \leq f(x) \leq f(y)$.

Accordingly D is inductive and so by **ZL** has a maximal element m. We claim that $f(m) = m$. To prove this it suffices to show that $f(m) \leq m$; since B is a base, for this it suffices in turn to prove that

(*) $$\forall b \in B[b \leq f(m) \Rightarrow b \leq m].$$

Since m is maximal in D, to prove (*) it clearly suffices to prove

$$\forall b \in B[b \leq f(m) \Rightarrow m \vee b \in D],$$

i.e.

$$\forall b \in B[b \leq f(m) \Rightarrow m \vee b \leq f(m \vee b)],$$

i.e.

(**) $$\forall b \in B[b \leq f(m) \Rightarrow m \vee b \leq f(m) \wedge f(b)].$$

So suppose $b \in B$ and $b \leq f(m)$. We already know that $m \leq f(m)$, and $m \leq f(b)$ follows from $b \leq f(m)$ and the self-adjointness of f. Thus $m \leq f(m) \wedge f(b)$. Also $b \leq f(m) \wedge f(b)$ since we are given $b \leq f(m)$ and f is inflationary on B. Hence $m \vee b \leq f(m) \wedge f(b)$ as required, and (**) follows. ■

If R be a binary relation on a set A, an *R-clique* in A is a subset U of A such that

$$\forall x \in A[x \in U \Leftrightarrow \forall y \in U.\ xRy]$$

The **Clique Property (CP)** is the assertion that, for any reflexive symmetric binary relation R, an R-clique exists.

Now we can show that **ZL, FP** *and* **CP** *are all constructively equivalent:*

ZL ⇒ FP has been established above.

FP ⇒ CP. Let R be a symmetric reflexive binary relation on a set A. Define the function F on the power set **P**A of A [1] to itself by $F(X) = \{y \in A: \forall x \in X.xRy\}$.The symmetry of R is tantamount to the self-adjointness of F and the reflexivity of R to the assertion that F is inflationary on the base $\{\{a\}: a \in A\}$ for **P**A. Accordingly **FP** yields a fixed point $U \in$ **P**A for F, that is, an R-clique in A.

CP ⇒ ZL. Let $(P, \leq)$ be a inductive poset, and define R to be the symmetric reflexive relation $x \leq y \vee y \leq x$ on P. **CP** yields an R-clique U in P, which is evidently a chain in P, and so, by the inductivity of P, has an upper bound u. We claim that u is a maximal element of P. For suppose $u \leq x$. Then clearly $\forall y \in U$. xRy, whence $x \in U$, and so $x \leq u$. Therefore $x = u$, and u is maximal. ■

The equivalence between **FP** and **CP** may be further explicated by the following observation. Let f be a self-adjoint map on a complete poset P which is inflationary on a set B of generators, and let R be the symmetric reflexive relation $x \leq f(y)$ on B. Then there are mutually inverse correspondences φ, ψ between the set F of fixed points of f (which is easily shown to coincide with the set of maximal elements of $\{x \in P: x \leq f(x)\}$) and the set $\mathscr{C}$ of R-cliques. These correspondences are given, respectively, by $\varphi(m) = \{x \in B: x \leq m\}$ for $m \in F$ and $\psi(X) = \bigvee X$ for $X \in \mathscr{C}$.

This relationship can be described in category-theoretic terms. Let $\mathcal{Rel}$ be the category whose objects are pairs (A, R) with

---

[1] Note that **P**A is a complete partially ordered set under inclusion.

R a reflexive symmetric relation on a set A, and with relation-preserving maps as arrows. Let $\mathcal{T}$ be the category whose objects are triples (P, B, f) with P a complete poset, B a subset of P, and f a self-adjoint map on P which is inflationary on B; an arrow $p: (P, B, f) \to (P', B', f')$ in $\mathcal{T}$ is a $\bigvee$-preserving map $P \to P'$ sending B into B' such that $p(f(x)) \leq f'(p(x))$ for all $x \in L$. We define the functors $F: \mathcal{Rel} \to \mathcal{T}$ and $G: \mathcal{T} \to \mathcal{Rel}$ as follows. Given $\mathbf{A} = (A, R)$ and $h: \mathbf{A} \to (A', R') = \mathbf{A'}$ in $\mathcal{Rel}$, we define $F\mathbf{A} = (\mathbf{P}A, \{\{a\} : a \in A\}, R^*)$ with $R^*(X) = \{y \in A : \forall x \in X.xRy\}$; and Fh: $F\mathbf{A} \to F\mathbf{A'}$ by $(Fh)(X) = \{h(x): x \in X\}$. Given $\mathbf{P} = (P, B, f)$ and p: $\mathbf{P} \to (P', B', f')$ in $\mathcal{T}$ we define $G\mathbf{L} = (B, f^{\sim})$, where $f^{\sim}$ is defined by $x\ f^{\sim}\ y$ iff $x \leq f(y)$ and Gp is the restriction of p to B.

Then F is left adjoint to G, and the unit of the adjunction is iso. So F is full and faithful, and thus $\mathcal{Rel}$ is, up to isomorphism, a full coreflective subcategory of $\mathcal{T}$. The objects $\mathbf{P} = (P, B, f)$ of $\mathcal{T}$ for which the counit arrow $FG\mathbf{P} \to \mathbf{P}$ is epic are precisely those in which B is a base for P: call such objects *based*. The adjunction $F \dashv G$ then restricts to one between $\mathcal{Rel}$ and $\mathcal{T}$'s full subcategory $\mathcal{T}^*$ of based objects. So $\mathcal{Rel}$ is also, up to isomorphism, a full coreflective subcategory of $\mathcal{T}^*$.

DOING WITHOUT AC: "POINTLESS" TOPOLOGY

Many representation theorems take the form of assertions to the effect that such-and-such an abstract structure is always isomorphic to a set-theoretic or topological realization of that structure. Probably the earliest example of this type of theorem is Cayley's theorem to the effect that every group is isomorphic to a group of permutations of a set. In this case the "representing" set coincides with the underlying set of the group, so that the representation demands nothing more than what was provided by set theory in the first place. As another example, consider the Lindenbaum-Tarski theorem that any complete atomic Boolean algebra is isomorphic to the power set Boolean algebra of a set. Here the representing set is the set of atoms of the given Boolean algebra, thus again rendering unnecessary the provision of "new" points beyond what was given. In particular, no use of **AC** is needed to prove these assertions.

The situation is quite otherwise, however, for those representation theorems whose proofs depend upon **AC** in an essential way. The earliest example of a representation theorem of this type is undoubtedly the Stone Representation Theorem for Boolean algebras to the effect that any Boolean algebra to the algebra of clopen subsets of a certain topological space—the *Stone space* of *B*. Here, the elements, or *points,* of the Stone space are the *ultrafilters* in *B*. Now while some of these (the so-called *principal* ultrafilters) may be identified with the elements of *B,* the proof of the theorem requires the presence of non-principal ultrafilters—new ideal "points" of *B* whose existence is entirely dependent on the applicability of the Boolean prime ideal theorem, and hence on **AC.** Another example is the Stone-Gelfand-Naimark representation of any *C**-algebra *A* as the ring of continuous real-valued functions on a compact Hausdorff space. Here the points of the representing space are the maximal ideals in *A,* whose existence, once again, depends on **AC.** Still another example is the Grothendieck representation of an arbitrary commutative ring *R* with identity as a ring of global sections of a sheaf of local rings over a compact T_0-space. Here the space is the *Zariski spectrum* of *R*: its points are the prime ideals in *R,* whose existence yet again depends on **AC.** While not strictly speaking a representation theorem, the Stone-Čech compactification theorem is of a similar nature, since the points of the Stone-Čech compactification of a given (completely regular Hausdorff) space X are the maximal ideals in the ring of (bounded) continuous real-valued functions on X. While some of these correspond to points of X, the majority do not, and their existence is entirely dependent on **AC.**

In each of these cases, then, a certain topological space is constructed, the existence of (the vast majority of) whose points depends on **AC**[1]**.** If one wants to avoid the use of **AC**—for example if one wants to work in a general topos-theoretic setting in which the Law of Excluded Middle is not affirmed—and yet at the same time retain as much of the content of these types of results as possible, it is natural to seek a formulation of topological ideas in a form that avoids all mention of "points". This has come to be known as "pointless topology".

The origins of pointless topology can be traced to the observation, originating with Ehresmann [1957] and Bénabou [1958] that the essential characteristics of a topological space are carried, not by its set of points, but by the complete Heyting algebra of its open sets. Thus complete Heyting algebras came to be regarded as "generalized topological spaces" in their own right. As "frames" these were studied by C. H. Dowker and D. Papert Strauss throughout the 1960s and 70s (see, e.g., their [1966]. 1966 and [1972]). Isbell [1972] observed that not the category of frames itself, but rather its *opposite*—whose objects he termed *locales*—was in fact the appropriate generalization of the category of topological spaces. Locales accordingly became known as "pointless" spaces and the study of the properties of the category of locales "pointless topology". The growth of topos theory, and

[1] Related examples include the identification of choice functions with extreme points and with points of Boolean fuzzy sets.

more particularly the study of sheaf toposes, greatly stimulated the development of pointless topology. It was Joyal who first observed that the notion of locale provides the correct concept of topological space within a topos (a view later exploited to great effect in Joyal and Tierney [1984]) and, more generally, in any context where **AC** is not available. This latter observation was strikingly confirmed by Johnstone [1981] who showed that Tychonoff's theorem that the product of compact spaces is compact, known to be equivalent to **AC**, can, suitably formulated in terms of locales, be proved without it. Johnstone became one of the champions of pointless topology, expounding the subject most persuasively in his book [1982], and elsewhere (e.g. in [1983a])[1].

Pointless topology rests on the concept of a *frame*, which is defined to be a complete lattice L satisfying the infinite distributive law

$$x \wedge \bigvee_{i \in I} y_i = \bigvee_{i \in I} x \wedge y_i \,.$$

It is easily shown that any frame is a Heyting algebra in which the $\Rightarrow$ operation is given by $a \Rightarrow b = \bigvee\{x\colon x \wedge a \leq b\}$. As examples of frames, we have:

- the open set lattice $\mathcal{O}(X)$ of a topological space X.
- the power set Boolean algebra $\mathbf{P}A$ of a set A.

---

[1] A "logical" approach to pointless topology—*formal spaces*—was introduced by Fourman and Grayson [1982]. Here the (constructive) theory of locales was developed in a logical framework using the concept of *intuitionistic propositional theory*. Each such theory was shown to engender (the dual of) a locale—its formal space—whose properties reflect those of the theory: in particular, semantic completeness of the theory (that is, possession of sufficient models for a completeness theorem to hold for it) was shown to correspond to the condition that the formal space be a genuine space (that is, possess enough points). Under the name *formal topology*, this approach has been considerably refined and developed by G. Sambin and his students and associates within the more demanding constructive framework of Martin-Löf type theory (see, e.g. Sambin [1988], Valentini [1996]).

- the frame Idl(D) of ideals of a distributive lattice D. Here Idl(D) is the set of all ideals of D, partially ordered by inclusion. In Idl(D), the meet of two ideals is given by their intersection, and the join of a family of ideals by the ideal generated by their union.

A *frame homomorphism* between frames L and L' is a map $f: L \to L'$ preserving finite meets and arbitrary joins. If $f: X \to Y$ is a continuous map of topological spaces, then the inverse map f^{-1}: $\mathcal{O}(Y) \to \mathcal{O}(X)$ is a frame homomorphism. The category $\mathcal{Frm}$ of frames is the category whose objects are frames and whose arrows are frame homomorphisms. The category $\mathcal{Loc}$ of *locales* is the opposite of the category of frames. The arrows of $\mathcal{Loc}$ are called *continuous maps*. We write $\mathcal{O}$ for the functor[1] $\mathcal{Top} \to \mathcal{Loc}$ which sends a space to its lattice of open sets and a continuous map $f: X \to Y$ to the function f^{-1}: $\mathcal{O}(Y) \to \mathcal{O}(X)$.

We now introduce the concept of a *point* of a locale. Since a point of a space X in the usual sense corresponds to a continuous map $1 \to X$, where 1 is the one point space, it is natural to define a *point* of a locale L to be a continuous map $\mathcal{O}(1) = 2 \to L$, i.e., a frame homomorphism p: $L \to 2$. Now it is easily seen that p is completely determined by $p^{-1}(0)$ or $p^{-1}(1)$, which are, respectively, a prime ideal and a prime filter in L. Now since p preseves arbitrary joins, $p^{-1}(0)$ must be a principal ideal, since $p(\bigvee(p^{-1}(0))) = 0$, so that $p^{-1}(0) = \{x: x \leq \bigvee(p^{-1}(0))\}$. Equivalently, $p^{-1}(1)$ must be a *completely prime* filter, i.e. it satisfies

$$\bigvee X \in p^{-1}(1) \Leftrightarrow \exists x \in X(x \in p^{-1}(1)).$$

[1] Here $\mathcal{Top}$ is the category of topological spaces as defined in Appendix II.

Call an element a of L *prime* if, for any $x, y \in L$, $x \wedge y \leq a \Rightarrow x$. Thus an element is prime if and only if it generates a prime principal ideal. Accordingly points of L correspond bijectively to prime elements of L, as well as to completely prime filters in L. Write $\pi(L)$ for the set of points of L.

Now define the map $\varphi: L \rightarrow \mathbf{P}(\pi(L))$[1] by taking $\varphi(a)$ to be the set of points $p: L \rightarrow 2$ such that $p(a) = 1$ (equivalently, the set of prime elements $x \in L$ such that $a \not\leq x$). It is easy to show that φ is a frame homomorphism, so that its image is a topology on $\pi(L)$. With this topology, $\pi(L)$ becomes a topological space called the *space of points* of L. The map φ will be regarded both as an arrow $L \rightarrow \mathcal{O}(\pi(L))$ in $\mathcal{Frm}$ and as a continuous map $\mathcal{O}(\pi(L)) \rightarrow L$ in $\mathcal{Loc}$. It can be shown that the assignment $L \mapsto \pi(L)$ defines a functor $\mathcal{Loc} \rightarrow \mathcal{Top}$ which is right adjoint to $\mathcal{O}$.

In general, the map $L \rightarrow \mathcal{O}(\pi(L))$, while obviously surjective, is not an isomorphism since it can fail to be injective. (Consider, for example, a complete Boolean algebra B regarded as a locale; the points of B may be identified with its atoms, so that the map φ sends $a \in B$ to the set of atoms x such that $x \leq a$. Thus φ is injective if and only if B is atomic.) In fact φ is an isomorphism of frames if and only if L satisfies the condition

$$\forall a \in L \forall b \in L[a \not\leq b \Rightarrow \exists p \in \pi(L)[p(a) = 1 \text{ and } p(b) = 0]].$$

or equivalently: for every a, b such that $a \not\leq b$ there is a prime element c such that $b \leq c$ but $a \not\leq c$, or a completely prime filter containing a but not b. A locale satisfying this condition is called *spatial* or said to *have enough points*. It is readily shown that a locale is spatial if and only if each element can be expressed as a meet of

[1] Recall that **P**X is the powerset of X.

prime elements. Obviously $\mathcal{O}(X)$ is spatial for every topological space X.

There are a number of conditions that can be placed on a locale to ensure that it is spatial (or possesses at least one point) but in every case the proof of this fact requires the use of **AC** (usually in the form of the existence of prime or maximal ideals) in furnishing the requisite points. For example, consider the condition of *coherence*. Let us call an element a of a complete lattice L *finite* if for every subset $A \subseteq L$ with $\bigvee A \geqslant a$, there exists a finite $F \subseteq X$ with $\bigvee A \geqslant a$. Then a locale L is said to be *coherent* if (i) every element is expressible as a join of finite elements and (ii) the finite elements of L form a sublattice of L. It can be shown that coherent locales are precisely those isomorphic to frames of the form Idl(D)[1], and it follows from this (together with **ZL**) that any coherent locale is spatial.

Here is a sketch of the proof. One first shows that the prime elements of Idl(D) are precisely the prime ideals of D. Then, to show that Idl(D) is spatial, it suffices to show that, if I, J are are ideals of D with $I \not\subseteq J$, there exists a prime ideal K of D with $J \subseteq K$, $I \not\subseteq K$. Let a be any element of $I - J$. An application of **ZL** yields an ideal K maximal with respect to the property of containing J and being disjoint from the filter $\{x: a \leq x\}$. It can then be shown that K is prime and so meets the requirements.

As another example, consider the condition on a locale corresponding to that of compactness of a topological space. Thus we say that a locale L is *compact* if its top element 1 is finite. Using **ZL** it is not hard to show that any nontrivial compact locale L has

[1] Johnstone [1982], 64.

at least one point. For by **ZL** L has a maximal ideal I, which is also prime. Since $1 \notin I$, it follows from compactness that $\bigvee I \neq 1$, so that the principal ideal $\{x: x \leq \bigvee I\}$ is proper. But this ideal evidently contains I and so is identical with I by maximality. Thus I is itself principal. Since I is also prime, it determines a point of I.

Not every compact locale is spatial. This can be seen by starting with a nonspatial locale L and adding a new top element to it: the resulting locale L' is easily seen to be compact and to have the same prime elements as L, so that L' is also nonspatial. But if one adds to compactness the condition on a locale corresponding to *regularity* of a topological space (that is, if there is a base of closed neighbourhoods at each point of the space), it turns out (assuming **ZL**) that locales satisfying the combined condition are spatial. How is regularity defined for a locale? Given a locale L, define the relation $\preccurlyeq$ on L by $a \preccurlyeq b$ iff $a^* \vee b = 1$. Clearly $a \preccurlyeq b \Rightarrow a \leq b$. We call L *regular* if, for any $a \in L$, $a = \bigvee\{b: b \preccurlyeq a\}$. When L is $\mathcal{O}(X)$, this condition says that every open set U can be covered by open subsets whose closures are contained in U; and this is readily seen to be equivalent to the usual definition of regularity for the space X.

Now let us sketch the proof that (assuming **ZL**), every compact regular locale L is spatial. Suppose that $a \nleq b$ in L. Then by the regularity of L, there is $c \preccurlyeq a$, i.e. $c^* \vee a = 1$, with $c \nleq b$. From the latter it follows that $b \vee c^* \neq 1$. Now consider the set $L' = \{x \in L: b \vee c^* \leq x\}$. With the order inherited from L, L' is then a nontrivial locale with bottom element $b \vee c^*$ and top element 1. Since L is compact, so is L'. Hence by the above result (whose proof uses **ZL**), L' has a prime principal filter I. It is now easily verified that $J = \{x \in L: x \vee b \vee c^* \in I\}$ is a prime principal filter in L

containing b but not a. Hence L is spatial.

A point of interest here is that Tychonov's theorem that the product of compact spaces is compact (which we have observed is actually equivalent to **AC**) has been formulated and proved in a "pointless" version for compact locales without the use of **AC.**[1] (Here the product of locales in $\mathcal{Loc}$ is actually the *coproduct* of frames in $\mathcal{Frm}$.)

Another natural candidate for "localeization" is the property of *local compactness,* i.e. the property that there is a base of compact neighbourhoods at each point). Given two elements a, b of a locale L, we define $a \trianglelefteq b$ to mean that, for any $A \subseteq L$, if $b \leq \bigvee A$, then $a \leq \bigvee F$ for some finite $F \subseteq A$. L is aid to be *locally compact* if, for any $b \in L$, $b = \bigvee\{a\colon a \trianglelefteq b\}$. It is not hard to show, that if X is locally compact, then $\mathcal{O}(X)$ is a locally compact locale, and that the converse holds when X is regular. Also, just as each compact regular space is locally compact, the same assertion holds for locales.

Using **ZL**, it can be shown that every locally compact locale is spatial. Here is a very rough sketch of the proof. Call a filter F in a locale L *open* if for any $a \in F$ there is $b \trianglelefteq a$ such that $b \in F$. Now let L be a locally compact locale. A straightforward argument shows that L has the *interpolation property,* namely, that if $a \trianglelefteq b$ in L, then there is $c \in L$ with $a \trianglelefteq c \trianglelefteq b$. Then one can employ the interpolation property inductively to show that, for any $a \ntrianglelefteq b$ in L, there is an open filter F containing a but not b. Next, **ZL** is used to enlarge F to an open filter M maximal amongst those not

[1] Johnstone [1981]. While Johnstone's proof does not use **AC**, it does require an application of transfinite induction and so cannot be regarded as being fully constructive. For compact *regular* locales, however, the use of transfinite induction can be avoided and the proof is fully constructive.

containing b. It can then be shown that M is completely prime, so we have found a completely prime filter containing a but not b. Thus L is spatial.

Finally, we mention another topological result which has been provided with a "pointless" formulation whose proof avoids the use of **AC**, namely, the *Stone-Čech compactification theorem*[1]. Here it becomes necessary to introduce for locales the condition corresponding to complete regularity of a topological space. To do this, one first defines a *scale* on a locale L to be a sequence of elements $(a_q: q \in \mathbb{Q} \cap [0, 1])$ such that $a_p \lessdot a_q$ whenever $p < q$. Then write $a \prec b$ if there exists a scale $(c_q: q \in \mathbb{Q} \cap [0, 1])$ such that $a \leq c_0$ and $c_1 \leq b$. The locale L is said to be *completely regular* if for every $b \in L$ we have $b = \bigvee\{a: a \prec b\}$. Now it can be shown that $a \prec b$ is equivalent to the condition that there exists a continuous map $f: L \rightarrow \mathcal{O}(\mathbb{R})$—that is, a frame homomorphism $\mathcal{O}(\mathbb{R}) \rightarrow L$ — for which $f((0, \infty)) \wedge a = 0_L$ and $f((-\infty, 1)) \leq b$. If we think of a and b as open sets in a topological space X, and f as a continuous real-valued function on X, this may be understood as expressing the condition "f takes values ≤ 0 inside a and $\geqslant 1$ outside b". Accordingly complete regularity of L corresponds to the condition "for every element x of a, there is a continuous real-valued function f on X such that $f(x) \leq 0$ and $f \geqslant 1$ outside a". This is precisely the usual condition of complete regularity for a topological space.

Banaschewski's and Mulvey's construction pivots on the locale βL of *completely regular* ideals of L, where an ideal I of L is completely regular if for any $a \in I$ there is $b \in I$ such that $a \prec b$. They show that βL is a compact completely regular locale, and that it has exactly the properties one would demand of a Stone-Čech compactification in the localic setting.

[1] Johnstone [1982], Banaschewski and Mulvey [1980].

IV
Consistency and Independence of the Axiom of Choice

In this chapter we give a necessarily compressed account of how **AC** is shown to be consistent with, and independent of, Zermelo-Fraenkel set theory.

ZERMELO-FRAENKEL SET THEORY

The *language of set theory* is a first-order language $\mathcal{L}$ with equality, which also includes a binary relation symbol $\in$ (membership). The individual variables x, y, z, are understood to range over *sets*, but we shall also permit the formation of *class terms* $\{x: \varphi(x)\}$ for each formula $\varphi(x)$. The term $\{x: \varphi(x)\}$ is understood to denote the class of all sets x such that $\varphi(x)$. We assume that classes satisfy the *Comprehension Principle:*

$$\forall y[y \in \{x: \varphi(x)\} \Leftrightarrow \varphi(y)].$$

We shall employ the standard set-theoretic abbreviations, such as $x \subseteq y$ for "x is included in y", $\varnothing$ for the empty set, $<x, y>$ for the ordered pair of x, y, $\bigcup x$ for the union of x, **P**x for the power set of x, $u \times v$ for the Cartesian product of u, v, " dom(u) for the domain of u, Fun(f) for "f is a function, etc. We also write V for the class of all sets, i.e. $\{x: x = x\}$.

Zermelo-Fraenkel set theory (**ZF**) is the theory in $\mathcal{L}$ based on the following axioms[1]:

- *Extensionality* $\forall x \forall y[\forall z(z \in x \Leftrightarrow z \in y) \Rightarrow x = y]$.
- *Separation* $\forall u \exists v \forall x[x \in v \Leftrightarrow x \in u \wedge \varphi(x)]$.

- *Pairing* $\forall x \forall y \exists u \forall z[z \in u \Leftrightarrow z = x \vee z = y]$.
- *Replacement* $\forall u[\forall x{\in}u \exists y\, \varphi(x, y) \Rightarrow \exists v \forall x{\in}u \exists y{\in}v\, \varphi(x, y)]$.
- *Union* $\forall u \exists v \forall x[x \in v \Leftrightarrow \exists y{\in}u(x \in y)]$.
- *Power set* $\forall u \exists v \forall x[x \in v \Leftrightarrow \forall y{\in}x(y \in u)]$.
- *Infinity* $\exists u[\varnothing \in u \wedge \forall x{\in}u \exists y{\in}u(x \in y)]$.
- *Regularity* $\forall u[u \neq \varnothing \Rightarrow \exists x{\in}u \forall y{\in}u(y \notin x)]$.

A class U or set u is *transitive* if $v \subseteq U$ (resp. $v \subseteq u$) whenever $v \in U$ (resp. $v \in u$). The *transitive closure* $\mathrm{TC}(x)$ of a set x is the least transitive set containing x, i.e. $\{x\} \cup x \cup \bigcup x \cup \bigcup\bigcup x \cup \ldots$. An *ordinal* is a transitive set which is well-ordered by the membership relation $\in$; we write Ord(x) for "x is an ordinal".We use letters α, β, γ,... for ordinals; we write $\alpha < \beta$ for $\alpha \in \beta$. The least infinite ordinal is denoted by ω. The class ORD of ordinals is then itself well-ordered by <, which makes it possible to define sets by *recursion* on the ordinals. In particular we define the sets V_α for $\alpha \in$ ORD by

$$V_\alpha = \{x: \exists \xi < \alpha[x \subseteq V_\xi]\}.$$

The axiom of regularity implies that each set x is a member of some V_α; the least such α is called the *rank* of x and written rank(x).

Let R be a relation, i.e. a class of ordered pairs. R is said to be *well-founded* if for each set u the class $\{x: xRu\}$ is a set and each nonempty set u has an element x such that yRx for no $y \in u$. If R is a well-founded relation, the *principle of induction* on R—which is provable in **ZF**—is the assertion

$$\forall x[\forall y(yRx \Rightarrow \varphi(y)) \Rightarrow \varphi(x)] \Rightarrow \forall x \varphi(x),$$

for an arbitrary formula $\varphi(x)$. The *principle of recursion* on R – which is also provable in **ZF** – is the assertion that if F is any class of ordered pairs defining a single-valued mapping of V into V (such a class is called a *function* on V and we write as usual $F: V \to V$) then there is a (unique) function $G: V \to V$ such that

$$\forall u[G(u) = F(\langle u, G \mid Ru \rangle)],$$

where $G \mid v$ is the restriction of G to v, i.e. $G \cap (u \times V)$.

The Axiom of Regularity implies that $\in$ is well-founded, and so we have as special cases the *principle of* $\in$*-induction*

$$\forall x[\forall y(y \in x \Rightarrow \varphi(y)) \Rightarrow \varphi(x)] \Rightarrow \forall x \varphi(x),$$

and $\in$*-recursion:* for any $F: V \to V$ there is $G: V \to V$ such that

$$\forall u[G(u) = F(\langle u, G \mid u \rangle)].$$

Again, the relation $\text{rank}(x) < \text{rank}(y)$ is well-founded and so we have the *principle of induction on rank:*

$$\forall x[\forall y(\text{rank}(y) < \text{rank}(x) \Rightarrow \varphi(y)) \Rightarrow \varphi(x)] \Rightarrow \forall x \varphi(x).$$

If U is a class, and σ is a sentence of $\mathcal{L}$, the *relativization* $\sigma^{(U)}$ of σ to U is the sentence obtained from σ by restricting all the quantifiers in σ to U, that is, replacing each existential quantifier $\exists x$ by $\exists x \in U$ and each universal quantifier $\forall x$ by $\forall x(x \in U \Rightarrow ...)$. The sentence $\sigma^{(U)}$ may be regarded as asserting that σ is true, or holds, in the structure $\mathfrak{U} = \langle U, \in \rangle$, or that the latter is a (class) *model* of σ. The **ZF** axioms may then be construed as asserting that the universal structure $\mathfrak{V} = \langle V, \in \rangle$ is a model of **ZF.**

THE RELATIVE CONSISTENCY OF AC

In $\mathcal{L}$ we may take the Axiom of Choice in the form **AC1**, i.e.

$$\forall u \exists f[\text{Fun}(f) \wedge \text{dom}(f) = u \wedge \forall x \in u[u \neq \varnothing \Rightarrow f(x) \in x]].$$

We write **ZFC** for **ZF + AC.**

In $\mathcal{L}$ **AC** can also be formulated in a *global* version, namely,

GAC *there is a function* $F: V \to V$ *such that, for all* $u \neq \varnothing$, $F(u) \in u$.

F is called a *global choice function*. Clearly **GAC** implies **AC.**

The idea behind Gödel's proof of consistency of **AC** relative to **ZF** is to "carve out" a class model $\mathfrak{U} = \langle U, \in \rangle$ of **ZF + GAC** from the universal structure $\mathfrak{V}$, which we have already observed is a model of **ZF**. This procedure will take place entirely within **ZF** in the sense that, for each axiom σ of **ZF + GAC** the sentence $\sigma^{(U)}$, i.e. the assertion that $\mathfrak{U}$ is a model of σ, is provable in **ZF.** It follows from this that **AC** is consistent relative to **ZF** in the sense that, if **ZF** is consistent, so is **ZF + GAC.**

As we remarked in Chapter I, Gödel's original proof of the consistency of **AC** used the concept of *constructible* set to obtain $\mathfrak{U}$. Here we shall sketch the simpler proof based on the concept of *ordinal definable* set.

Informally, a set a is ordinal definable if it is definable from some finite set of ordinals, i.e. if there is a property $P(y_1, ..., y_n, x)$ and ordinals $\alpha_1, ..., \alpha_n$ such that, for any x, $P(\alpha_1, ..., \alpha_n, x) \Leftrightarrow x = a$. The formal counterpart of this concept within $\mathcal{L}$ is "definable within some structure $\mathfrak{V}_\alpha = \langle V_\alpha, \in \rangle$". Thus write $D(u)$ for the term in $\mathcal{L}$ representing the set of all subsets of u which are first-order definable in the structure $(u, \in)$ [1]. Now we can define

$$OD(x) \equiv \exists\alpha[x \in D(V_\alpha)].$$

The class $OD = \{x: OD(x)\}$ is the class of *ordinal definable sets*. It can then be proved[2] in **ZF** that, for any formula $\varphi(y, x, z_1, ..., z_n)$,

$$(*) \qquad \forall x[\exists\alpha_1 \dots \exists\alpha_n \forall y[\varphi(y, x, \alpha_1, ..., \alpha_n) \Leftrightarrow x = y] \Rightarrow OD(x)$$

[1] Such a term can be constructed within $\mathcal{L}$; see e.g. Bell and Machover [1977] or Kunen [1980].

[2] See, e.g. Bell and Machover [1977] or Kunen [1980].

This shows that the informal definition of ordinal definability implies the formal version. It follows in particular that *every ordinal is ordinal definable,* i.e. ORD $\subseteq$ OD.

It follows from (*) that OD is *closed under definability* in the sense that, if $\tau(x_1, ..., x_n)$ is any term of $\mathcal{L}$, then

(**) $$a_0, \dots, a_n \in \text{OD} \Rightarrow \tau(a_0, \dots, a_n) \in \text{OD}.$$

Next, it can be shown that OD has a *definable well-ordering.* By this we mean that one can construct a formula $\psi(x, y)$ for which the formal statement "ψ defines a (strict) well-ordering of OD" is a theorem of **ZF.** This is done[1] by defining terns s_1 and s_2 of $\mathcal{L}$ by

$s_1(x)$ = *least ordinal* α *such that* $x \in \text{D}(V_\alpha)$ if $x \in$ OD; 0 *otherwise.*

$s_1(x)$ = *least natural number n such that n is the code number of a formula defining x in* $\mathfrak{V}_{s_1(x)}$; 0 *otherwise.*

The formula $\psi(x, y)$ is then defined by

$$\psi(x, y) \equiv \text{OD}(x) \wedge \text{OD}(y) \wedge [\, s_1(x) < s_1(y) \vee [s_1(x) = s_1(y) \wedge s_2(x) < s_2(y)]].$$

We shall write $x \prec y$ for $\psi(x, y)$ and call $\prec$ the *definable well-ordering of* OD.

Unfortunately, OD cannot be proved to be transitive and so cannot be shown to be a model of **ZF.** To remedy this we replace OD with the class HOD of *hereditarily ordinal definable sets,* i.e. those $x \in$ OD such that all members of x, members of members of x, etc., are in OD. Formally, we define

$$\text{HOD}(x) \equiv \text{TC}(x) \subseteq \text{OD} \qquad \text{HOD} = \{x: \text{HOD}(x)\}.$$

[1] For details see, e.g., Bell and Machover [1977], Kunen [1980].

It is easy to show that (i) ORD $\subseteq$ HOD $\subseteq$ OD; (ii) HOD is transitive; (iii) for any set a, if $a \in$ OD and $a \subseteq$ HOD, then $a \in$ HOD.

One can now prove in **ZF** the

Theorem. $\mathfrak{HOD} = \langle \text{HOD}, \in \rangle$ *is a model of* **ZF** +**GAC.**

Proof. Extensionality holds in $\mathfrak{HOD}$ since HOD is transitive and Regularity holds since it holds in any class.

For Separation, note that if $u \in$ HOD, then, for any formula $\varphi(x)$, $\{x \in u: \varphi^{(\text{HOD})}(x)\} \in$ HOD.

The Axioms of Pairing, Union, Replacement and Power set in $\mathfrak{HOD}$ all assert that HOD contains "large enough sets". Each is proved in a similar way, using (**), and (i) - (iii) above. For example, in the case of Power Set, let $u \in$ HOD and define $\mathbf{P}^*u$ = $\mathbf{P}u \cap$ HOD. Obviously $\mathbf{P}^*u \subseteq$ HOD, and (**) implies that Also $\mathbf{P}^*u \in$ OD. Hence $\mathbf{P}^*u \in$ HOD and so Power Set holds in $\mathfrak{HOD}$.

The Axiom of Infinity holds in $\mathfrak{HOD}$ since $\omega \in$ HOD.

Finally, to show that **GAC** holds in $\mathfrak{HOD}$, observe that the definable well-ordering $\prec$ of OD restricts to a well-ordering $\prec'$ of HOD. Then we can define a global choice function F : HOD $\rightarrow$ HOD by setting, for each nonempty $u \in$ HOD, $F(u) = \prec'$ - least element of u. ■

By the remarks above, it follows that, **GAC**, and hence also **AC**, is relatively consistent with **ZF.**

THE INDEPENDENCE OF AC

The method of proving the independence of **AC** outlined here is known as the method of *Boolean-valued models.* This was developed in the 1960s by Robert Solovay and Dana Scott, building on Cohen's original technique of forcing[1].

To describe the method of Boolean-valued models of set theory, we need to introduce the idea of a Boolean-valued structure. Let B be a complete Boolean algebra. A *B-valued structure* to be a triple $\mathfrak{S} = \langle S, [\![\bullet = \bullet]\!], [\![\bullet \in \bullet]\!]\rangle$, where S s a class and $[\![\bullet = \bullet]\!]$, $[\![\bullet \in \bullet]\!]$ are maps $S \times S \to B$ satisfying the conditions

$$[\![u = v]\!] = 1_B$$
$$[\![u = v]\!] = [\![v = u]\!]$$
$$[\![u = v]\!] \wedge [\![v = w]\!] \leq [\![u = w]\!]$$
$$[\![u = v]\!] \wedge [\![u \in w]\!] \leq [\![v \in w]\!]$$
$$[\![v = w]\!] \wedge [\![u \in v]\!] \leq [\![u \in w]\!].$$

for $u, v, w \in S$.

Let $\mathcal{L}(S)$ be the language obtained from $\mathcal{L}$ by adding a name for each element of S. For convenience we identify each element of S with its name in $\mathcal{L}(S)$ and use the same symbol for both. The maps $[\![\bullet = \bullet]\!]$, $[\![\bullet \in \bullet]\!]$ can be extended to a map $[\![\bullet]\!]$ defined on the class of all $\mathcal{L}(S)$-sentences recursively via:

$$[\![\sigma \wedge \tau]\!] = [\![\sigma]\!] \wedge [\![\tau]\!]$$
$$[\![\sigma \vee \tau]\!] = [\![\sigma]\!] \vee [\![\tau]\!]$$
$$[\![\sigma \Rightarrow \tau]\!] = [\![\sigma]\!] \Rightarrow [\![\tau]\!]$$
$$[\![\neg\sigma]\!] = [\![\sigma]\!]^*$$
$$[\![\exists x\varphi(x)]\!] = \bigvee_{u \in S} [\![\varphi(u)]\!]$$
$$[\![\forall x\varphi(x)]\!] = \bigwedge_{u \in S} [\![\varphi(u)]\!]$$

For each sentence σ, $[\![\sigma]\!] \in B$ is called the *truth value* of σ in $\mathfrak{S}$; σ is *true* in $\mathfrak{S}$, written $\mathfrak{S} \models \sigma$, if $[\![\sigma]\!] = 1_B$ and *false* in $\mathfrak{S}$ if $[\![\sigma]\!] = 0_B$. $\mathfrak{S}$

[1] For a systematic account of Boolean-valued models, see Bell [2005].

is a *(Boolean-valued) model* of a set T of $\mathcal{L}(S)$-sentences if each member of T is true in $\mathfrak{S}$. It is not hard to show that, if $\mathfrak{S}$ is a model of T, and $T \vdash \sigma$, then $\mathfrak{S} \models \sigma$.

Now the idea is to build, for each complete Boolean algebra B, a B-valued structure $\mathfrak{V}^{(B)}$ called the *(full) universe of B-valued* sets or the *B-extension of the universe of sets,* which can be proved, in **ZFC,** to be itself a Boolean-valued model of **ZFC.** It follows that any sentence σ which is *false* in some $\mathfrak{V}^{(B)}$ must be *independent* of **ZFC.** By selecting B with finesse, the independence of numerous set-theoretic principles, such as the Axiom of Constructibility and the Continuum Hypothesis can in particular be established using this method.

We observe that full universes of Boolean-valued sets cannot be used for the purpose of demonstrating the independence of **AC** from **ZF,** since it is a theorem of **ZFC** that **AC** is always true in any $\mathfrak{V}^{(B)}$. To obtain a Boolean-valued model of **ZF** in which **AC** is *false,* $\mathfrak{V}^{(B)}$ must be replaced with a *submodel* associated with the action of a certain type of group. Such submodels are the Boolean-valued analogues of Fraenkel's symmetric models mentioned in Chapter I. We defer discussion of these until later.

Now suppose given a complete Boolean algebra B which we assume is a *set,* i.e. $B \in V$. The *class* $V^{(B)}$ *of B-valued sets* is defined as follows. First, we define by recursion the sets $V_\alpha^{(B)}$ for each ordinal α:

$$V_\alpha^{(B)} = \{x : \mathrm{Fun}(x) \wedge \mathrm{range}(x) \subseteq B \wedge \exists \xi < \alpha[\mathrm{domain}(x) \subseteq V_\xi^{(B)}\}.$$

Then we define

$$V^{(B)} = \{x : \exists \alpha[x \in V_\alpha^{(B)}\}.$$

It is easily seen that a B-valued set is precisely a B-valued function whose domain is a set of B-valued sets. We write $\mathcal{L}^{(B)}$ for the language $\mathcal{L}(V^{(B)})$.

The basic principle for establishing facts about B-valued sets is the

Induction Principle for $V^{(B)}$. *For any formula* $\varphi(x)$, *if*

$$\forall x \in V^{(B)}[\forall y \in \mathrm{dom}(x)\ \varphi(y) \Rightarrow \varphi(x)],$$

then

$$\forall x \in V^{(B)}\ \varphi(x).$$

This is easily proved by induction on rank.

We now proceed to turn $V^{(B)}$ into a B-valued structure. This is done by defining $[\![u = v]\!]^{(B)}$ and $[\![u \in v]\!]^{(B)}$ by the equations:

$$[\![u \in v]\!]^{(B)} = \bigvee_{y \in dom(v)} [v(y) \wedge [\![u = y]\!]^{(B)}]$$

$$[\![u = v]\!]^{(B)} = \bigwedge_{x \in dom(u)} [u(x) \Rightarrow [\![x \in v]\!]^{(B)}] \wedge \bigwedge_{y \in dom(v)} [v(y) \Rightarrow [\![y \in u]\!]^{(B)}] \quad .$$

These can be justified by recursion on a certain well-founded relation[1].

It can now be shown by $\in$-induction that $\mathfrak{B}^{(B)} = \langle V^{(B)}, [\![\bullet \in \bullet]\!]^{(B)}, [\![\bullet = \bullet]\!]^{(B)} \rangle$ is a B-valued structure. This structure is called the *universe of B-valued sets.* We assume that $[\![\bullet]\!]^{(B)}$ has been extended to the class of all $\mathcal{L}^{(B)}$ - sentences as above: we shall usually omit the superscript $^{(B)}$.

Of help in calculating truth values in $\mathfrak{B}^{(B)}$ are the rules:

$$[\![\exists x \in u\varphi(x)]\!]^{(B)} = \bigvee_{x \in \mathrm{dom}(u)} [\![\varphi(u)]\!]^{(B)} \qquad [\![\forall x\varphi(x)]\!]^{(B)} = \bigwedge_{x \in \mathrm{dom}(u)} [\![\varphi(x)]\!]^{(B)} .$$

There is a natural map ^: $V \to V^{(B)}$ defined by $\in$-recursion as follows:

$$x^\wedge = \{\langle y^\wedge, 1_B \rangle : y \in x\}.$$

It is then easily shown that, for $x \in V$, $u \in V^{(B)}$,

---

[1] Bell [2005], p. 23

$$[\![u \in x\hat{}]\!]^{(B)} = \bigvee_{y \in x} [\![u = y\hat{}]\!]^{(B)}.$$

Now in **ZFC** it can be shown that $\mathfrak{B}^{(B)}$ *is a B-valued model of* **ZFC.** We verify the Axiom of Separation and **AC** in $\mathfrak{B}^{(B)}$.

We recall that the Axiom of Separation is the scheme

$$\forall u \exists v \forall x[x \in v \Leftrightarrow x \in u \wedge \varphi(x)].$$

To see that each instance is true in $\mathfrak{B}^{(B)}$, let $u \in V^{(B)}$, define $v \in V^{(B)}$ by dom(v) = dom(u) and, for $x \in$ dom(v), $v(x) = u(x) \wedge [\![\varphi(x)]\!]$.

Then

$[\![\forall x[x \in v \Leftrightarrow x \in u \wedge \varphi(x)]\!] = [\![\forall x \in v[x \in u \wedge \varphi(x)]]\!] \wedge [\![\forall x \in u[\varphi(x) \Rightarrow x$

Now

$$[\![\forall x \in v[x \in u \wedge \varphi(x)]]\!] = \bigwedge_{x \in \text{dom}(v)} [[u(x) \wedge [\![\varphi(x)]\!]] \Rightarrow [x \in u \wedge [\![\varphi(x)]\!]]] = 1_B.$$

Similarly

$$[\![\forall x \in u[\varphi(x) \Rightarrow x \in v]\!] = 1_B$$

and the assertion follows.

As for **AC**, we sketch a verification in $\mathfrak{B}^{(B)}$ of the equivalent Ordinal Covering Principle (Chapter III). We recall that this is

$$\forall u \exists \alpha \exists f[\text{Fun}(f) \wedge \text{dom}(f) = \alpha \wedge u \subseteq \text{range}(f).$$

To establish its truth in $\mathfrak{B}^{(B)}$, take any $u \in V^{(B)}$; **AC** implies that there is an ordinal α and a function g of α onto domain(u). Define $f \in V^{(B)}$ by

$$f = \{<\beta\hat{}, g(\beta)>^{(B)} : \beta < \alpha\} \times \{1_B\}$$ [1]

It is easy to show that $[\![\text{Ord}(\alpha\hat{})]\!] = 1_B$, so it suffices to show that

$$\mathfrak{B}^{(B)} \models \text{Fun}(f) \wedge \text{dom}(f) = \alpha\hat{} \wedge u \subseteq \text{range}(f).$$

We verify that $\mathfrak{B}^{(B)} \models \text{dom}(f) = \alpha\hat{}$. For we have, for $x \in V^{(B)}$,

[1] Here $<u,v>^{(B)}$ is the *B-ordered pair in* $V^{(B)}$, that is, the *B*-set playing the role of the ordered pair in $V^{(B)}$. If we define $\{u\}^{(B)} = \{<u, 1_B>\}$ and $\{u, v\}^{(B)} = \{<u, 1_B>, <v, 1_B>\}$, then $<u,v>^{(B)}$ may be defined as $\{\{u\}^{(B)}, \{u, v\}^{(B)}\}^{(B)}$.

$$\llbracket \exists y[< x, y > \in f] \rrbracket = \bigvee_{z \in V^{(B)}} \llbracket < x, z > \in f \rrbracket$$
$$= \bigvee_{z \in V^{(B)}} \bigvee_{\beta < \alpha} \llbracket \beta\hat{} = x \rrbracket \wedge \llbracket g(\beta) = z \rrbracket$$
$$= \bigvee_{\beta < \alpha} \llbracket \beta\hat{} = x \rrbracket \wedge \bigvee_{z \in V^{(B)}} \llbracket g(\beta) = z \rrbracket$$
$$= \bigvee_{\beta < \alpha} \llbracket \beta\hat{} = x \rrbracket$$
$$= \llbracket x \in \alpha\hat{} \rrbracket.$$

The remaining conjunctions are similarly verified.

The complete Boolean algebras normally employed in formulating independence proofs are the *regular open algebras* associated with topological spaces. If X is a topological space, a subset U is said to be *regular open* if $U = \overset{\circ}{\overline{U}}$, that is, if U coincides with the interior of its closure. The family $\mathsf{R}(X)$ of all regular open subsets of X forms a complete Boolean algebra under the partial order of inclusion, in which $\bigvee_{i \in I} U_i = \overset{\circ}{\overline{\bigcup_{i \in I} U_i}}$, $\bigwedge_{i \in I} U_i = \overset{\circ}{\bigcap_{i \in I} U_i}$, and $U^* = X - \overline{U}$. $\mathsf{R}(X)$ is called the *regular open algebra of* X.

Now let X and Y be nonempty sets, where Y has at least 2 elements. Write $C(X, Y)$ for the set of all mappings with domain a finite subset of X and range a subset of Y. Partially order $C(X, Y)$ by inverse inclusion and write $(P, \leq)$ for the resulting poset. For $p \in P$ let

$$N(p) = \{f \in Y^X : p \subseteq f\}.$$

Subsets of Y^X of the form $N(p)$ form a base for the product topology on Y^X , when Y is assigned the product topology. Each $N(p)$ is then a clopen (closed-and-open) subset of Y^X in this topology. In particular, each $N(p)$ is a regular open subset of Y^X, and it is easy to verify that the map $p \mapsto N(p)$ is an order-isomorphism of P onto a dense subset of $B = \mathrm{R}(Y^X)$. (Here a subset A of a Boolean algebra B is *dense* if $0 \notin A$ and for any $x \in B$ such that $x \neq 0_B$ there is $a \in A$ for which $a \leq x$.) We agree to identify p and $N(p)$, so that P may be regarded as a dense subset of B. We also agree to use $\leq$ for the partial ordering on B.

We now turn to the construction of the submodels of $\mathfrak{V}^{(B)}$ in which **AC** can be falsified. For this we require the concept of a group action on a class. Thus let G be a group, and X a class. An *action* of G on X is a map $<g, x> \mapsto g \cdot x$: $G \times X \to X$ satisfying $1 \cdot x = x$, $(gh) \cdot x = g \cdot (h \cdot x)$. (We shall usually write gx for $g \cdot x$.) Under these conditions we say that G *acts* on X. For each $g \in G$, the map π_g: $X \to X$ defined by $\pi_g(x) = g \cdot x$ is a permutation of X.

If B is a Boolean algebra, by an action of G on B we mean an action of G on B by automorphisms, that is, one in which each π_g is an automorphism of B.

We extend the notion of group action to Boolean-valued structures by defining an *action* of a group G on a B-valued structure $\mathfrak{S} = <S, [\![\bullet \in \bullet]\!], [\![\bullet = \bullet]\!]>$ to be a pair of actions of G on B and on S satisfying

$$[\![gu = gv]\!] = g \cdot [\![u = v]\!] \qquad [\![gu \in gv]\!] = g \cdot [\![u \in v]\!].$$

It is easily shown by induction on complexity of formulas that, for any formula $\varphi(x_1, ..., x_n)$ of $\mathcal{L}$, any $u_1, ..., u_n \in S$, and any $g \in G$,

$$g \cdot [\![\varphi(u_1, ..., u_n)]\!] = [\![\varphi(gu_1, ..., gu_n)]\!].$$

Now let G be a group acting on the complete Boolean algebra B. Define the map $<g, x> \mapsto gx$: $G \times V^{(B)} \to V^{(B)}$ by recursion on the well-founded relation $y \in \mathrm{dom}(x)$ via:

$$gu = \{<gx, g \cdot u(x)>: x \in \mathrm{dom}(u)\}.$$

It can then be shown that this defines an action of G on $\mathfrak{V}^{(B)}$ such that (i) for any $u \in V^{(B)}$, $g \in G$, $\mathrm{dom}(gu) = \{gx: x \in \mathrm{dom}(u)\}$ and for any $x \in \mathrm{dom}(u)$, $(gu)(gx) = g \cdot u(x)$; and (ii) for any $x \in V$, $gx^\wedge = x^\wedge$.

Here is a sketch of how the independence of **AC** is proved. Let Ω be the group of all permutations of ω and for each $n \in \omega$ let $\Omega_n = \{g \in \Omega: gn = n\}$. We choose a certain complete Boolean algebra B and construct a certain subclass V^* of $V^{(B)}$ such that

(i) V^* is the underlying class of a B-valued model $\mathfrak{V}^*$ of **ZF;**

(ii) $x^\wedge \in V^*$ for all $x \in V$;

(iii) Ω acts on $\mathfrak{V}^*$;

(iv) for each $x \in V^*$, there is a finite subset J of ω (called a *support* of x) such that $gx = x$ for every $g \in \bigcap_{n \in J} \Omega_n = \Omega_J$;

(v) there is an infinite "set of distinct reals" $s = \{u_n: n \in \omega\}$ in V^* such that $gu_n = u_{gn}$ for all $g \in \Omega$ and $n \in \omega$.

From this it will follow that, in $\mathfrak{V}^*$, s is infinite but s has no denumerable subset, so *a fortiori* **AC** fails in $\mathfrak{V}^*$. For suppose f is any map (in V^*) of $\omega^\wedge$ [1] into s. Then, by (iv), f has a finite support J. If f were injective, then there would be $n \notin J$ such that $u_n \in \mathrm{range}(f)$. Choose $n' \notin \{n\} \cup J$ and let $g \in \Omega$ be the permutation of ω which interchanges n and n' but leaves the remaining integers undisturbed. If $u_n = fm^\wedge$, then $u_{n'} = u_{gn} = gu_n = g(fm^\wedge) = (gf)(gm^\wedge) = fm^\wedge = u_n$, contradicting $u_n \neq u_{n'}$. Hence, in $\mathfrak{V}^*$, s does not have a denumerable subset, so **AC** fails there.

---

[1] $\omega^\wedge$ plays the role of ω both in $\mathfrak{V}^{(B)}$ and in $\mathfrak{V}^{(\Gamma)}$.

For each $x \in V^{(B)}$ define stab(x), the *stabilizer* of x, to be the set $\{g \in \Omega: gx = x\}$. Then it follows from condition (iv) that, for each $x \in V^*$, stab (x) is a member of the *filter of subgroups* of Ω generated by the Ω_n, i.e. the family Γ of all subgroups of Ω which contain at least one Ω_J. This leads to the idea of considering an *arbitrary filter of subgroups* of an *arbitrary group* G. Also, since we want Ω to act on $\mathfrak{V}^*$, we must have $x \in V^* \Rightarrow gx \in V^* \Rightarrow$ stab(gx) $\in \Gamma$. But it is easy to verify that stab(gx) = g stab(x) g^{-1}, so we shall want Γ to satisfy $H \in \Gamma \Rightarrow gHg^{-1} \in \Gamma$. Under these conditions Γ is said to be *normal*. Finally, we shall write $V^{(\Gamma)}$, $\mathfrak{V}^{(\Gamma)}$ in place of V^*, $\mathfrak{V}^*$ to indicate the dependence of the construction on Γ.

Thus let G be a group acting on the complete Boolean algebra B and let Γ be a filter of subgroups of G. That is, Γ is a nonempty set of subgroups of G such that (a) $H, K \in \Gamma \Rightarrow H \cap K \in \Gamma$, (b) $H \in \Gamma$ and $H \subseteq K$, K a subgroup of $G \Rightarrow K \in \Gamma$. Γ is *normal* if $g \in G$ and $H \in \Gamma \Rightarrow gHg^{-1} \in \Gamma$.

We know that G acts on $\mathfrak{V}^{(B)}$; so for each $x \in V^{(B)}$ we can define the *stabilizer* stab(x) by stab(x) = $\{g \in \Omega: gx = x\}$. Clearly stab(x) is a subgroup of G. By analogy with the definition of the $V_\alpha^{(B)}$, we define the sets $V_\alpha^{(\Gamma)}$ recursively as follows:

$$V_\alpha^{(\Gamma)} = \{x : \text{Fun}(x) \wedge \text{range}(x) \subseteq B \wedge \text{stab}(x) \in \Gamma \wedge \exists\xi < \alpha[\text{domain}(x) \subseteq V_\xi^{(\Gamma)}\}.$$

Then we define

$$V^{(\Gamma)} = \{x : \exists\alpha[x \in V_\alpha^{(\Gamma)}\}.$$

Clearly $V^{(\Gamma)} \subseteq V^{(B)}$, and

$x \in V^{(\Gamma)} \Leftrightarrow Fun(x) \wedge \text{range}(x) \subseteq B \wedge \text{domain}(x) \subseteq V^{(\Gamma)} \wedge \text{stab}(x) \in \Gamma.$

For $u, v \in V^{(\Gamma)}$, we define $[\![u = v]\!]^{(\Gamma)}$ and $[\![u \in v]\!]^{(\Gamma)}$ recursively as we defined $[\![u = v]\!]^{(B)}$ and $[\![u \in v]\!]^{(B)}$, that is:

$$[\![u \in v]\!]^{(\Gamma)} = \bigvee_{y \in \text{dom}(v)} [v(y) \wedge [\![u = y]\!]^{(\Gamma)}]$$

$$[\![u = v]\!]^{(\Gamma)} = \bigwedge_{x \in \text{dom}(u)} [u(x) \Rightarrow [\![x \in v]\!]^{(\Gamma)}] \wedge \bigwedge_{y \in \text{dom}(v)} [v(y) \Rightarrow [\![y \in u]\!]^{(\Gamma)}] \ .$$

It is then readily shown by induction that $[\![u = v]\!]^{(\Gamma)} = [\![u = v]\!]^{(B)}$ and $[\![u \in v]\!]^{(\Gamma)} = [\![u \in v]\!]^{(B)}$, so that $\mathfrak{V}^{(\Gamma)} = \langle V^{(\Gamma)}, [\![\bullet = \bullet]\!]^{(\Gamma)}, [\![\bullet \in \bullet]\!]^{(\Gamma)} \rangle$ is a B-valued structure. We denote by $\mathcal{L}^{(\Gamma)}$ the language for $\mathfrak{V}^{(\Gamma)}$, that is, the result of expunging from $\mathcal{L}^{(B)}$ all constant symbols not denoting elements of $V^{(\Gamma)}$. For any sentence σ of $\mathcal{L}^{(\Gamma)}$, we write $[\![\sigma]\!]^{(\Gamma)}$ for the truth value of σ in $\mathfrak{V}^{(\Gamma)}$.

The following two facts are readily established by induction: (i) *for any* $x \in V$, we have $x^\wedge \in V^{(\Gamma)}$; (ii) *G acts on* $\mathfrak{V}^{(\Gamma)}$. It follows from (ii) that, for any for any formula $\varphi(x_1, ..., x_n)$ of $\mathcal{L}$, any $u_1, ..., u_n \in V^{(\Gamma)}$, and any $g \in G$,

$$g \cdot [\![\varphi(u_1, \ldots, u_n)]\!]^{(\Gamma)} = [\![\varphi(gu_1, \ldots, gu_n)]\!]^{(\Gamma)}.$$

One can now prove in **ZF** the

Theorem. $\mathfrak{V}^{(\Gamma)}$ *is a model of* **ZF.**

This is proved in a way similar to that for the analogous result for $\mathfrak{V}^{(B)}$. As before, we verify Separation. Thus let $\varphi(x, v_1, ..., v_n)$ be an $\mathcal{L}$-formula and let $u, a_1, ..., a_n \in V^{(\Gamma)}$. Define $v \in V^{(B)}$ by $\text{dom}(v) = \text{dom}(u)$ and

$$v(x) = u(x) \wedge [\![\varphi(x, a_1, \ldots, a_n]\!]^{(\Gamma)}.$$

It now suffices to show that $v \in V^{(\Gamma)}$, for then it is readily verified, as for $V^{(B)}$, that

$$\mathfrak{B}^{(\Gamma)} \models \forall x[x \in v \Leftrightarrow x \in u \wedge \varphi(x, a_1, ..., a_n).$$

Since $\text{dom}(v) = \text{dom}(u) \subseteq V^{(\Gamma)}$, to show that $v \in V^{(\Gamma)}$ it is enough to show that $\text{stab}(v) \in \Gamma$. And since $\text{stab}(u)$, $\text{stab}(a_1)$, ..., $\text{stab}(a_n)$ are all in Γ and Γ is a filter, it will be enough to show that

(*) $\quad A = \text{stab}(u \cap \text{stab}(a_1) \cap ... \cap \text{stab}(a_n) \subseteq \text{stab}(v).$

If $g \in A$, then $\text{dom}(gv) = \{gx: x \in \text{dom}(v)\} = \{gx: x \in \text{dom}(u)\} = \text{dom}(gu) = \text{dom}(u) = \text{dom}(v)$. Also, if $x \in \text{dom}(v)$, then $x = gy$ with $y \in \text{dom}(u)$, so that

$$\begin{aligned}(gv)(x) &= (gv)(gy)\\ &= g \cdot v(y)\\ &= g \cdot u(y) \wedge [\![\varphi(gy, ga_1, \ldots, ga_n)]\!]^{(\Gamma)}\\ &= (gu)(gy) \wedge [\![\varphi(x, a_1, \ldots, a_n)]\!]^{(\Gamma)}\\ &= u(x) \wedge [\![\varphi(x, a_1, \ldots, a_n)]\!]^{(\Gamma)}\\ &= v(x).\end{aligned}$$

Hence $gv = v$ and $g \in \text{stab}(v)$. This proves (*) and Separation in $\mathfrak{B}^{(\Gamma)}$.

We shall specify B, G, and Γ so that $\mathfrak{B}^{(\Gamma)} \models \neg$**AC.** This will establish the independence of **AC** from **ZF.**

Let P be the poset $C(\omega \times \omega, 2)$, partially ordered by inverse inclusion, let X be the product space $2^{\omega\times\omega}$, and let B be the regular open algebra $\mathsf{R}(X)$ of X. Then, as observed above, P may be regarded as a dense subset of B when each $p \in P$ is identified with the element $N(p) = \{f \in 2^{\omega \times \omega}: p \subseteq f\}$ of B.

Again let Ω be the group of permutations of ω. Ω can be made to act on B in the following way. Each $g \in \Omega$ induces a homeomorphism g^* of X onto itself via

$$(g^*f)<m, n> = f<m, gn>.$$

We define the action $<g, b> \mapsto gb$ of Ω on B by

$$gb = g^{*-1}[b] = \{f \in X: g^*f \in b\}.$$

For each $n \in \omega$ let Ω_n be the subgroup $\{g \in \Omega: gn = n\}$ and let Γ be the filter of subgroups of Ω generated by the Ω_n, that is, the set of subgroups of Ω containing at least one subgroup $\bigcap_{n \in J} \Omega_n = \Omega_J$ for finite $J \subseteq \omega$. It is readily verified that Γ is normal.

We shall need a

Lemma. If $p \in P$, J is a finite subset of ω and $n \notin J$, then there is $g \in \Omega_J$ such that $p \wedge gp \neq 0_B$ and $gn \neq n$.

Proof. Take $n' \notin J \cup \{n\}$ so that $<m, n'> \notin \mathrm{dom}(p)$ for any m (possible, since J and dom(p) are finite) and let $g \in \Omega$ be the permutation of ω which interchanges n and n' but leaves the remaining integers undisturbed. Then certainly $g \in \Omega_J$ and $gn \neq n$. To verify that $p \wedge gp \neq 0_B$, recall that p has been identified with $N(p)$ and observe that

$$\begin{aligned} g \cdot N(p) &= \{f \in 2^{\omega\times\omega} : p \subseteq g^* f\} \\ &= \{f \in 2^{\omega\times\omega} : <i, j> \in \mathrm{dom}(p) \Rightarrow f <i, gj> = p <i, j>\}. \end{aligned}$$

Let $i_1, ..., i_k$ be a list of the i such that $<i, n> \in \mathrm{dom}(p)$. Then

$$\begin{aligned} p \wedge gp &= N(p) \cap g \cdot N(p) \\ &= \{f \in 2^{\omega\times\omega} : p \subseteq f \text{ and } f <i_j, n'> \; = p <i_j, n> \text{ for } j = 1, \ldots, k\} \\ &\neq \varnothing, \end{aligned}$$

since $<i_j, n'> \notin \mathrm{dom}\,(p)$ for $j = 1, ..., k$. ■

Finally we prove the

Theorem. $\mathfrak{V}^{(\Gamma)}$ is a model of $\neg$**AC**.

Proof. To prove this it will be convenient to employ the *forcing relation*[1] $\Vdash$ between P and the class of sentences of $\mathcal{L}^{(\Gamma)}$. This is defined by

$$p \Vdash \sigma \text{ iff } p \leq [\![\sigma]\!]^{(\Gamma)}.$$

Clearly $[\![\sigma]\!]^{(\Gamma)} = 0_B$ iff $p \Vdash \sigma$ for no $p \in P$, and $[\![\sigma]\!]^{(\Gamma)} = 1_B$ iff $p \Vdash \sigma$ for all $p \in P$. Clearly, also, $\Vdash$ is *persistent* in the sense that, if $p \Vdash \sigma$ and $q \leq p$, then $q \Vdash \sigma$. Two further, easily established facts about $\Vdash$ we shall need are that (i) $p \Vdash \neg\sigma$ if and only if $q \nVdash \sigma$ for all $q \leq p$, so that $p \nVdash \neg\sigma$ iff there is $q \leq p$ for which $q \Vdash \sigma$ and (ii) for $a \in V$, $p \Vdash \exists x \in a^\wedge \varphi(x)$ iff there exist $q \leq p$ and $x \in a$ such that $q \Vdash \varphi(x^\wedge)$. These facts will be used below without comment.

For each $n \in \omega$ define $u_n \in B^{\mathrm{dom}(\omega^\wedge)}$ by

$$u_n(m^\wedge) = \{h \in 2^{\omega\times\omega} : h<m,n> = 1\}.$$

It is then easily verified that $\mathfrak{B}^{(\Gamma)} \models u_n \in \omega^\wedge$ and that $\mathfrak{B}^{(\Gamma)} \models u_n \neq u_{n'}$ for $n \neq n'$. We next establish:

(1) $gu_n = u_{gn}$. For clearly we have $\mathrm{dom}(gu_n) = \mathrm{dom}(u_{gn})$. Also, for $m \in \omega$,

$$\begin{aligned}(gu_n)m^\wedge &= (gu_n)gm^\wedge \\ &= g \cdot u_n(m^\wedge) \\ &= g^{*-1}[\{h \in 2^{\omega\times\omega} : h<m,n>=1\}] \\ &= \{h \in 2^{\omega\times\omega} : g^*h<m,n>=1\} \\ &= \{h \in 2^{\omega\times\omega} : h<m,gn>=1\} \\ &= u_{gn}(m^\wedge),\end{aligned}$$

whence (1).

[1] For a full account of forcing see Bell [2005] or Kunen [1980].

It follows immediately from (1) that $\Omega_n \subseteq \mathrm{stab}(u_n)$, so $\mathrm{stab}(u_n) \in \Gamma$ and $u_n \in V^{(\Gamma)}$.

Now put $s = \{u_n: n \in \omega\} \times 1_B$; then $gs = s$ for any $g \in \Omega$, so $s \in V^{(\Gamma)}$. Since $\mathfrak{B}^{(\Gamma)} \models u_n \neq u_{n'}$ for $n \neq n'$, it follows that

$$\mathfrak{B}^{(\Gamma)} \models s \textit{ is infinite.}$$

We claim that

$$\mathfrak{B}^{(\Gamma)} \models s \textit{ has no denumerable subset,}$$

which will prove the theorem. To establish the claim, it suffices to show that, for each $f \in V^{(\Gamma)}$,

$$[\![\mathrm{Fun}(f) \wedge f \textit{ is injective} \wedge \mathrm{dom}(f) = \omega\hat{} \wedge \mathrm{range}(f) \subseteq s]\!] = 0_B.$$

And to prove this it suffices to show that for no $p_0 \in P$ is it the case that

$$p_0 \Vdash \mathrm{Fun}(f) \wedge f \textit{ is injective} \wedge \mathrm{dom}(f) = \omega\hat{} \wedge \mathrm{range}(f) \subseteq s.$$

Suppose on the contrary that (*) held for some p_0. We shall find $q \leq p_0$ such that $q \Vdash \neg\mathrm{Fun}(f)$, in violation of the persistence of $\Vdash$, so yielding the required contradiction.

We first observe that

(2) $\quad p \Vdash x \in s$ iff $\forall q \leq p \exists r \leq q \exists n \in \omega [r \Vdash x = u_n]$.

For we have

$$p \Vdash x \in s \text{ iff } p \leq \bigvee_{n\in\omega} [\![x = u_n]\!]$$

$$\text{iff } p \wedge \bigwedge_{n\in\omega} [\![x \neq u_n]\!] = 0$$

$$\text{iff } \forall q \leq p \ [q \nleq \bigwedge_{n\in\omega} [\![x \neq u_n]\!]]$$

$$\text{iff } \forall q \leq p \ \exists n \in \omega \ [q \nleq [\![x \neq u_n]\!]]$$

$$\text{iff } \forall q \leq p \ \exists n \in \omega \ [q \nVdash x \neq u_n]$$

$$\text{iff } \forall q \le p\, \exists n \in \omega \exists r \le q\, [r \Vdash x = u_n].$$

Now since $f \in V^{(\Gamma)}$ it has a finite support J, i.e. there is a finite subset $J \subseteq \omega$ such that $\Omega_J \subseteq \text{stab}(f)$. Let $J = \{n_1, ..., n_j\}$. Since $p_0 \Vdash f$ is injective $\wedge$ Fun(f), it follows that

$$p_0 \Vdash \exists x \in \omega\hat{}\,[f(x) \neq u_{n_1} \wedge \ldots \wedge f(x) \neq u_{n_j},$$

so that there is $p \le p_0$ and $m \in \omega$ such that

(3) $$p \Vdash f(m\hat{}) \neq u_{n_1} \wedge \ldots \wedge f(m\hat{}) \neq u_{n_j}.$$

Since $p_0 \Vdash f(m\hat{}) \in s$, so that $p \Vdash f(m\hat{}) \in s$, by (2) there are $r \le p$ and $n \in \omega$ such that

(4) $$r \Vdash f(m\hat{}) = u_n.$$

But from (3) we deduce

$$r \Vdash f(m\hat{}) \neq u_{n_1} \wedge \ldots \wedge f(m\hat{}) \neq u_{n_j},$$

and this, together with (4) implies $n \notin J$. By the above Lemma there is $g \in \Omega_J$ such that $r \wedge gr \neq 0$ and $gn \neq n$. It follows from (4) that

$$gr \Vdash (gf)(gm\hat{}) = gu_n.$$

But this, together with (1) and the fact that $g \in \Omega_J \subseteq \text{stab}(f)$ gives

$$gr \Vdash f(m\hat{}) = u_{gn}.$$

Since $r \wedge gr \neq 0_B$, there is $q \in P$ such that $q \le r$ and $q \le gr$. Then $q \le p_0$ and

$$q \Vdash f(m\hat{}) = u_n \wedge f(m\hat{}) = u_{gn}.$$

But from $gn \neq n$ it follows that $[\![u_{gn} \neq u_n]\!] = 1_B$, so that $q \Vdash u_{gn} \neq u_n$. Therefore $q \Vdash \neg$Fun(f), and the proof is complete. ■

V
The Axiom of Choice and Intuitionistic Logic

AC AND LOGIC

An initial connection between **AC** and logic can be discerned by returning to its formulation **AC3** in terms of relations, namely: any binary relation contains a function with the same domain. This version of **AC** is naturally expressible within a many-sorted second-order language $\mathscr{L}$ with individual variables $x, y, z, \ldots$, constant symbols **a, b, c,** ... function variables $f, g, h, \ldots$ and function symbols **f, g, h, ...** We assume that each individual variable x and each constant symbol **a** is assigned an (individual) sort **A,** indicated by writing x:**A** or **a:A** and that each function symbol f and each function symbol **f** is assigned a pair of sorts **A, B,** indicated by writing $f: \mathbf{A} \to \mathbf{B}$ or $\mathbf{f}: \mathbf{A} \to \mathbf{B}$. In either case, if x:**A** or **a:A,** then fx, f**a, f**x and **fa** are all terms of sort **B.**

In $\mathscr{L}$, binary relations are represented by formulas $\varphi(x, y)$ with two free individual variables x:**A**, y:**B**. The counterpart in $\mathscr{L}$ of the assertion **AC3** is then

ACL $\forall x{:}\mathbf{A}\, \exists y{:}\mathbf{B}\, \varphi(x, y) \Rightarrow \exists f{:}\mathbf{A} \to \mathbf{B} \forall x{:}\mathbf{A}\, \varphi(x, fx).$

This scheme of sentences is the standard logical form of **AC**.

Zermelo's original form of the Axiom of Choice, **AC1,** can be expressed as a scheme of sentences within a third-order language $\mathscr{L}^*$ extending $\mathscr{L}$. Accordingly we suppose $\mathscr{L}^*$ to contain in addition predicate variables $X, Y, Z, \ldots$ predicate constants **U, V, W, ...,** second-order function variables $F, G, H, \ldots$and second-order function constants **F, G, H, ...** . Predicate variables and constants are assigned *power* sorts of the form **PA,** where **A** is an individual sort, indicated by X:**PA** or **U:PA.** In either case, for x:**A** or **a:A,** $X(x)$, $X(\mathbf{a})$, $U(x)$ and $X(\mathbf{a})$ are all well-formed statements.

Each function variable or constant is assigned a pair of sorts **PA, B**, indicated by F: **PA → B** or **F: PA → B.** In either case, for X:**PA** or U:**PA,** FX, F**U, F**X, **FU** are all terms of sort **B.**

The scheme of sentences

AC1L $\forall X$:**PA** [$\Phi(X) \Rightarrow \exists x$:**A** $X(x)$] $\Rightarrow$

$\exists F$:**PA→A** $\forall X$:**PA** [$\Phi(X) \Rightarrow X(FX)$]

where $\Phi(X)$ is any formula containing at most the free variable X is the direct counterpart of **AC1** in $\mathscr{L}^*$.

Up to now we have tacitly assumed our background logic to be the usual classical logic. But the true depth of the connection between **AC** and logic emerges only when *intuitionistic* or *constructive* logic is brought into the picture. It is a remarkable fact that, assuming only the framework of intuitionistic logic together with certain mild further presuppositions, **AC** can be shown to yield the cardinal rule of classical logic, the Law of Excluded Middle **(LEM)**—the assertion that $p \vee \neg p$ for any proposition p. We shall first show that **LEM** can be derived, using the rules of intuitionistic logic, within $\mathscr{L}$ from **ACL** conjoined with the following additional principles:

- **Binary Sort Principle** There is a sort **2** and constants **0:2, 1:2** subject to the axioms $\mathbf{0} \neq \mathbf{1}$ and $\forall x$:**2**[$x = \mathbf{0} \vee x = \mathbf{1}$]
- **Binary Quotient Principle.** Call a formula $\rho(x$:**A**$, y$:**A**$)$ an *equivalence relation on* **A** if it satisfies the usual conditions of reflexivity, symmetry and transitivity. The Binary Quotient Principle is the assertion that for any equivalence relation ρ on **2** there is a sort **2/ρ** and two constants $\mathbf{0}_\rho$, $\mathbf{1}_\rho$ of sort **2/ρ** subject to the axioms (A) $\forall u$:**2/ρ** [$u = \mathbf{0}_\rho \vee u = \mathbf{1}_\rho$] and (B) $\mathbf{0}_\rho = \mathbf{1}_\rho \Leftrightarrow \rho(\mathbf{0}, \mathbf{1})$. Thus **2/ρ** represents the quotient of **2** by the equivalence relation ρ.

Now assume **ACL** and the two principles above. Given a sentence p, define the equivalence relation ρ on **2** by $\rho(x,y) \equiv (x = y \vee p)$. Note that then from Axiom (B) it follows that

(*) $$\mathbf{0}_\rho = \mathbf{1}_\rho \Leftrightarrow p.$$

Let $\varphi(u{:}\mathbf{2/\rho}, x{:}\mathbf{2})$ be the formula $(u = \mathbf{0}_\rho \wedge x = \mathbf{0}) \vee (u = \mathbf{1}_\rho \wedge x = \mathbf{1})$. From axiom (A) of the Binary Quotient Principle we infer $\forall u{:}\mathbf{2/\rho}\ \exists x{:}\mathbf{2}\ \varphi(u, x)$, so by **ACL** we can introduce a function symbol $\mathbf{f}\colon \mathbf{2/\rho} \to \mathbf{2}$ for which $\forall u{:}\mathbf{2/\rho}\ \varphi(u, \mathbf{f}u)$. It follows that $\varphi(\mathbf{0}_\rho, \mathbf{f}(\mathbf{0}_\rho)) \wedge \varphi(\mathbf{1}_\rho, \mathbf{f}(\mathbf{1}_\rho))$, which is equivalent to the conjunction of the two formulas

(a) $$\mathbf{f}(\mathbf{0}_\rho) = \mathbf{0} \vee [\mathbf{0}_\rho = \mathbf{1}_\rho \wedge \mathbf{f}(\mathbf{0}_\rho) = \mathbf{1}]$$

(b) $$[\mathbf{0}_\rho = \mathbf{1}_\rho \wedge \mathbf{f}(\mathbf{1}_\rho) = \mathbf{0}] \vee \mathbf{f}(\mathbf{1}_\rho) = \mathbf{1}$$

From (*) it follows that **(a)** implies $p \vee \mathbf{f}(\mathbf{0}_\rho) = \mathbf{0}$ and **(b)** implies $p \vee \mathbf{f}(\mathbf{1}_\rho) = \mathbf{1}$. Taking the conjunction of these and applying the distributive law gives

(**) $$p \vee [\mathbf{f}(\mathbf{0}_\rho) = \mathbf{0} \wedge \mathbf{f}(\mathbf{1}_\rho) = \mathbf{1}].$$

Now from $[\mathbf{f}(\mathbf{0}_\rho) = \mathbf{0} \wedge \mathbf{f}(\mathbf{1}_\rho) = \mathbf{1}]$ (and $\mathbf{0} \neq \mathbf{1}$) we deduce $\mathbf{f}(\mathbf{0}_\rho) \neq \mathbf{f}(\mathbf{1}_\rho)$, so (**) gives

(***) $$p \vee \mathbf{f}(\mathbf{0}_\rho) \neq \mathbf{f}(\mathbf{1}_\rho).$$

But $p \Rightarrow \mathbf{0}_\rho = \mathbf{1}_\rho \Rightarrow \mathbf{f}(\mathbf{0}_\rho) = \mathbf{f}(\mathbf{1}_\rho)$, so that $\mathbf{f}(\mathbf{0}_\rho) \neq \mathbf{f}(\mathbf{1}_\rho) \Rightarrow \neg p$. So it follows from (***) that $p \vee \neg p$, i.e. **LEM.**

Next, we show that **LEM** can be derived, using the rules of intuitionistic logic, within $\mathscr{L}^*$ from **AC1L** conjoined with the Binary Sort Principle and the following additional principles:

- **Predicative Comprehension Principle**
  $\exists X{:}\mathbf{PA}\ \forall x{:}\mathbf{A}[X(x) \Leftrightarrow \varphi(x)]$, where φ has at most the free variable x and contains no bound function or predicate variables.

- **Principle of Extensionality of Functions**
  $\forall F\mathbf{:PA \to A}\ \forall X\mathbf{:A}\ \forall Y\mathbf{:A}\ [X \approx Y \Rightarrow FX = FY]$, where $X \approx Y$ is an abbreviation for $\forall x\mathbf{:A}[X(x) \Leftrightarrow Y(x)]$, that is, X and Y are *extensionally equivalent*.

Now let p be a sentence. By Predicative Comprehension and Binary Sort, we may introduce predicate constants **U:P2**, **V:P2** together with the assertions

(1) $\forall x\mathbf{:2}[\mathbf{U}(x) \Leftrightarrow (p \vee x = 0)] \quad \forall x\mathbf{:2}[\mathbf{V}(x) \Leftrightarrow (p \vee x = 1)]$

Let $\Phi(X\mathbf{:P2})$ be the formula $X \approx U \vee X \approx V$. Then clearly we may assert $\forall X\mathbf{:P2}\ \exists x\mathbf{:2}\ [\Phi(X) \Rightarrow X(x)]$ so **AC1L** may be invoked to assert $\exists F\mathbf{:P2 \to 2}\ \forall X[\Phi(X) \Rightarrow X(FX))$. Now we can introduce a function constant **K** together with the assertion

(2) $$\forall X[\Phi(X) \Rightarrow X(\mathbf{K}X)].$$

Since evidently we may assert $\Phi(\mathbf{U})$ and $\Phi(\mathbf{V})$, it follows from (2) that we may assert $\mathbf{U}(\mathbf{KU})$ and $\mathbf{V}(\mathbf{KV})$, whence also, using (1),

$$[p \vee \mathbf{KU} = 0] \wedge [p \vee \mathbf{KV} = 1].$$

Using the distributive law, it follows that we may assert

$$p \vee [\mathbf{KU} = 0 \wedge \mathbf{KV} = 1].$$

From the presupposition that $0 \neq 1$ it follows that

(3) $$p \vee \mathbf{KU} \neq \mathbf{KV}$$

is assertable. But it follows from (1) that we may assert $p \Rightarrow \mathbf{U} \approx \mathbf{V}$, and so also, using Extensionality of Functions, $p \Rightarrow \mathbf{KU} = \mathbf{KV}$. This yields the assertability of $\mathbf{KU} \neq \mathbf{KV} \Rightarrow \neg p$, which, together with (3) in turn yields the assertability of

$$p \vee \neg p,$$

that is, **LEM**.

The fact that **AC** implies **LEM** seems at first sight to be at variance with the fact that **AC** taken as a *valid* principle in certain systems of constructive mathematics governed by intuitionistic logic, e.g. Bishop's Constructive Analysis[1] and Martin-Löf's Constructive Type Theory[2], but in which at the same time **LEM** is not affirmed.

Some light may be shed on the difficulty by observing that, in deriving **LEM** from **ACL** essential use was made of the Binary Quotient Principle and, in deriving **LEM** from **ACL1** similar use was made of both the Principles of Predicative Comprehension and Extensionality of Functions. It follows that, in systems of constructive mathematics affirming **AC** but not **LEM**, Constructive Type Theory for instance[3], *the Binary Quotient Principle and either the Predicative Comprehension Principle or the Principle of Extensionality of Functions must fail.*

Several observations concerning these facts should be made. It is a basic tenet of Constructive Type Theory that, to be able to assert that an object a has a specified property φ, one must be in possession of a *proof* that such is the case. So, on *a priori* grounds, the Predicative Comprehension Principle is not justified in Constructive Type Theory because, in attempting to replace a property φ by an extensionally equivalent predicate or set **U**, it cannot be guaranteed that whenever has **U(a)** one also has φ**(a),** since evidence for the first assertion does not necessarily produce a proof of the second. The Principle of Extensionality of Functions is not affirmable in Constructive Type Theory for essentially the same reason, namely that (using set-theoretic language) the value of a function defined on a (sub)set X depends not only on the variable member x of X but also on the *proof* that x is in fact in X. Thus suppose given sets A, B and a subset $X = \{x: \beta(x)\}$ of A. Write $d \vdash \alpha$ for "d is a proof of α". Then since **AC** holds in Constructive

---

[1] See Bishop and Bridges [1985].
[2] **AC** is actually *provable* in Constructive Type Theory: See Chapter VII below.
[3] See Chapter VII below.

Type Theory, from $\forall x \in A[\beta(x) \Rightarrow \exists y \in B\varphi(x, y)]$ we can infer the existence of a function f: $\{(x, p): p \vdash \beta(x)\} \Rightarrow B$ for which $\forall x \forall p[p \vdash \beta(x) \Rightarrow \varphi(x, f(x,p))]$. Given all this, let us attempt to derive **LEM** from **AC1L**. Here A is **P2**, the power set of 2 (supposing that to be present), $\beta(x)$ is $\exists x.\ x \in X$ (X a variable of sort P2), B is 2 and $\varphi(X, y)$ is $y \in X$. Now, given a sentence p, define the subsets U and V as were **U** and **V** above. Constructively, the only proof of $\exists x.\ x \in U$ available is to exhibit a member of U, and, since α is not known to be true, the sole exhibitable member of U is 0. Similarly, the only exhibitable member of V is 1. Writing $a = f(U, 0)$ and $b = f(V, 1)$, we derive the counterpart of (1) above as from Predicative Comprehension as before. But now while $p \Rightarrow U = V$, we cannot infer that $U = V \Rightarrow a = b$, so blocking the derivation of $p \Rightarrow a = b$.

Another way of looking at this is to observe that functions on predicates are given *intensionally*, and satisfy just the corresponding *Principle of Intensionality*, which may be stated as $\forall X\ \forall Y\ \forall F[X = Y \Rightarrow FX = FY]$. While this is essentially tautological, and so immune to failure, its extensional counterpart —the Principle of Extensionality—can easily be made to fail. Consider, for example, the predicates **P**: *rational featherless biped* and **Q**: *human being* and the function **K** on predicates which assigns to each predicate the number of words in its description. It is evident that **P** ≈ **Q** but **KP** = 3 and **KQ** = 2.

As for the Binary Quotient Principle, one notes that in Constructive Type Theory the conditions for affirming an identity statement $a = b$ are such as not to allow, as is permissible in set theory, automatic conversion of assertions of equivalence into assertions of identity of "equivalence classes". This is the case even for equivalence relations on two-element sets, so that the

Binary Quotient Principle is inadmissible within Constructive Type Theory.

In intuitionistic set theory (that is, set theory based on intuitionistic rather than classical logic: see below) both the Principles of Predicative Comprehension and Extensionality of Functions hold[1] and so there **AC** implies **LEM**[2]. *This means that adding* **AC** *to intuitionistic set theory* "tips it over" *into classical set theory.* This is the true "logical significance"of **AC**, at least as regards set theory.

Now what about **ZL**? In Chapter VI, we shall show that **ZL** has *no* nonconstructive purely logical consequences, and so in particular, unlike **AC**, cannot imply **LEM**. It follows that the derivation of **AC** from **ZL** in classical set theory cannot go through in intuitionistic set theory. Let us look into the matter[3].

Typically, applications of **ZL** take the following form. Suppose, for example, one wishes to show that a function possessing a certain property P exists with domain a certain set A. To do this one proves first that the collection $\mathcal{F}$ of functions with property P and and domain a subset of A is closed under unions of chains and then infers from **ZL** that $\mathcal{F}$ has a maximal element m. Finally a "one-step extension" argument is formulated so as to yield the conclusion that the domain of m is A itself. This "one-step" argument can be distilled into the *extension principle for* $\mathcal{F}$, namely

EP($\mathcal{F}$) $\qquad \forall f \in \mathcal{F} \forall x \in A \exists g \in \mathcal{F} [f \subseteq g \wedge x \in \text{domain}(g)]$.

[1] Here the predicate variables should be construed as variables ranging over sets.
[2] But in weak set theories lacking the axiom of extensionality the derivation of Excluded Middle from **AC** does not go through: some form of extensionality, or the existence of quotient sets for equivalence relations, needs to be assumed. See below.
[3] Bell [1995].

Applying this to the maximal m immediately yields the desired conclusion A = domain(m).

Now consider the derivation of **AC3** from **ZL** as indicated in Chapter II. A moment's thought reveals that, in terms of the extension principle as just stated, the relevant collection of functions $\mathscr{F}$ is the set $R^{\#}$ of subfunctions of a given relation R with domain A and codomain B, and the extended function g figuring in **EP**($\mathscr{F}$) is obtained from the given function f and the given element $x \in A$ by means of a classical definition by cases:

$g = f$ if $x \in$ domain(f), $g = f \cup <x, y>$ for some $y \in B$ such that $< x, y > \in R$ if $x \notin$ domain(f).

Moreover, if we write **EP** for the statement

$\forall R[R$ *is a binary relation* $\Rightarrow$ **EP**($R^{\#}$),

then the implication **ZL + EP ⇒ AC** is, plainly, constructively valid. It follows that **EP** must itself be nonconstructive. And indeed we can show that **EP** *implies* **LEM**.

To prove this in intuitionistic set theory, let 2 = {0, 1} and, given any proposition p, define $U = \{x \in 2: x = 0 \vee p\}$, $V = \{x \in 2: x = 1 \vee p\}$ and $R = (\{U\} \times U) \cup (\{V\} \times V)$. Then the function $f_0 = \{<U, 0>\}$ is in $R^{\#}$ and so **EP** yields a a function g in $R^{\#}$ extending f_0 such that domain(g) = $\{U, V\}$. Thus $g(U) = 0$ and $g(V) \in V$, so that $g(V) = 1 \vee p$. But clearly $p \Rightarrow V = U \Rightarrow g(V) = g(U) = 0$. Thus $g(V) \neq 0 \Rightarrow \neg p$, whence $g(V) = 1 \Rightarrow \neg p$. From this and $g(V) = 1 \vee p$ we conclude that $\neg p \vee p$, i.e. **LEM**.

CHOICE PRINCIPLES IN INTUITIONISTIC SET THEORY

As we have seen, in intuitionistic set theory **LEM** is derivable from **AC.** We are now going to show that each of a number of classically correct, but intuitionistically invalid logical principles, including **LEM** for sentences, is, in intuitionistic set theory, equivalent to a suitably *weakened* version of **AC**. Thus each of these logical principles may be viewed as a choice principle.

The system **IST** of intuitionistic set theory we shall work in is an intuitionistic theory formulated in the first-order language of set theory L introduced at the beginning of Chapter IV and based on the following axioms also stated there: *Extensionality, Separation, Pairing, Union,* and *Power Set.*

Let us begin by fixing some notation. For each set A we write PA for the power set of A, and QX for the set of *inhabited* subsets of A, that is, of subsets X of A for which $\exists x\ (x \in A)$. The set of functions from A to B is denoted by B^A; the class of functions with domain A is denoted by Fun(A). The empty set is denoted by 0, {0} by 1, and {0, 1} by 2.

We tabulate the following *logical schemes*[1]:

LEM	$\varphi \vee \neg\varphi$
SLEM	$\alpha \vee \neg\alpha$ (α any sentence)
Lin	$(\alpha \Rightarrow \beta) \vee (\beta \Rightarrow \alpha)$ (α, β any sentences)
Stone	$\neg\alpha \vee \neg\neg\alpha$ (α any sentence)
Ex	$\exists x[\exists x\alpha(x) \Rightarrow \alpha(x)]$ ($\alpha(x)$ any formula with at most x free)
Un	$\exists x[\alpha(x) \Rightarrow \forall x\alpha(x)]$ ($\alpha(x)$ any formula with at most x free)
Dis	$\forall x[\alpha \vee \beta(x)] \Rightarrow \alpha \vee \forall x\beta(x)$ (α any sentence, $\beta(x)$ any formula with at most x free)

[1] In addition to these logical schemes there is also the scheme—called by Lawvere and Rosebrugh [2003] the *higher dual distributive law*—

HDDL $\forall x[\alpha(x) \vee \beta(x)] \Rightarrow \exists x\alpha(x) \vee \forall x\beta(x)$.

It is not difficult to show that, over intuitionistic predicate logic, **HDDL** is equivalent to **Dis.**

Over intuitionistic logic, **Lin, Stone** and **Ex** are consequences of **SLEM**; and **Un** implies **Dis**. All of these schemes follow, of course, from **LEM,** the full Law of Excluded Middle.

We formulate the following *choice principles*—here X is an arbitrary set and $\varphi(x, y)$ an arbitrary formula of the language of **IST** with at most the free variables x, y:

$\mathbf{AC}_X$	$\forall x \in X\, \exists y\, \varphi(x,y) \Rightarrow \exists f \in \mathrm{Fun}(X)\, \forall x \in X\, \varphi(x,fx)$
$\mathbf{AC}^*_X$	$\exists f \in \mathrm{Fun}(X)\, [\forall x \in X\, \exists y\, \varphi(x,y) \Rightarrow \forall x \in X\, \varphi(x,fx)]$
$\mathbf{DAC}_X$	$\forall f \in \mathrm{Fun}(X)\, \exists x \in X\, \varphi(x,fx) \Rightarrow \exists x \in X\, \forall y\, \varphi(x,y)$
$\mathbf{DAC}^*_X$	$\exists f \in \mathrm{Fun}(X)\, [\exists x \in X\, \varphi(x,fx) \Rightarrow \exists x \in X\, \forall y\, \varphi(x,y)]$

The first two of these are forms of the Axiom of Choice for X; while classically equivalent, in **IST** $\mathbf{AC}^*_X$ implies $\mathbf{AC}_X$, but not conversely. The principles $\mathbf{DAC}_X$ and $\mathbf{DAC}^*_X$ are *dual* forms of the Axiom of Choice for X: classically they are both equivalent to $\mathbf{AC}_X$ and $\mathbf{AC}^*_X$, but in **IST** $\mathbf{DAC}^*_X$ implies $\mathbf{DAC}_X$, and not conversely.

We also formulate what we shall call the *weak extensional selection principle*, in which $\alpha(x)$ and $\beta(x)$ are any formulas with at most the variable x free:

WESP $\quad \exists x \in 2\alpha(x) \wedge \exists x \in 2\beta(x) \Rightarrow \exists x \in 2 \exists y \in 2[\alpha(x) \wedge \beta(y) \wedge [\forall x \in 2\beta(x)] \Rightarrow x = y]].$

This principle asserts that, for any pair of instantiated properties of members of 2, instances may be assigned to the properties in a manner that depends just on their extensions. **WESP** is a straightforward consequence of $\mathbf{AC}_{Q2}$. For taking $\varphi(u, y)$ to be $y \in u$ in $\mathbf{AC}_{Q2}$ yields the existence of a function f with domain $Q2$ such that $fu \in u$ for every $u \in Q2$. Given formulas $\alpha(x)$, $\beta(x)$, and assuming the antecedent of **WESP,** the sets $U = \{x \in 2\colon \alpha(x)\}$ and $V = \{x \in 2\colon \beta(x)\}$ are members of $Q2$, so that $a = fU \in U$, and

$b = fV \in V$, whence $\alpha(a)$ and $\beta(b)$. Also, if $\forall x{\in}2[\alpha(x) \Leftrightarrow \beta(x)]$, then $U = V$, whence $a = b$; it follows then that the consequent of **WESP** holds.

We are going to show that each of the logical principles tabulated above is equivalent (over **IST**) to a choice principle. Starting at the top of the list, we have first:

- **WESP** *and* **SLEM** *are equivalent over* **IST**.

Proof. Assume **WESP.** Let σ be any sentence and define

$$\alpha(x) \equiv x = 0 \vee \sigma \qquad \beta(x) \equiv x = 1 \vee \sigma\,.$$

With these instances of α and β the antecedent of **WESP** is clearly satisfied, so that there exist members a, b of 2 for which (1) $\alpha(a) \wedge \beta(b)$ and (2) $\forall x\ [[\forall x{\in}2[\alpha(x) \Leftrightarrow \beta(x)] \Rightarrow a = b$. It follows from (1) that $\sigma \vee (a = 0 \wedge b = 1)$, whence (3) $\sigma \vee a \neq b$. And since clearly $\sigma \Rightarrow \forall x{\in}2[\alpha(x) \Leftrightarrow \beta(x)]$ we deduce from (2) that $\sigma \Rightarrow a = b$, whence $a \neq b \Rightarrow \neg\sigma$. Putting this last together with (3) yields $\sigma \vee \neg\sigma$, and **SLEM** follows.

For the converse, we argue informally. Suppose that **SLEM** holds. Assuming the antecedent of **WESP**, choose $a \in 2$ for which $\alpha(a)$. Now (using **SLEM**) define an element $b \in 2$ as follows. If $\forall x{\in}2[\alpha(x) \Leftrightarrow \beta(x)]$ holds, let $b = a$; if not, choose b so that $\beta(b)$. It is now easy to see that a and b satisfy $\alpha(a) \wedge \beta(b) \wedge [\forall x{\in}2[\alpha(x) \Leftrightarrow \beta(x)] \Rightarrow a = b]$. **WESP** follows. ■

Next, we observe that, while $\mathbf{AC_1}$ is (trivially) provable in **IST,** by contrast

- $\mathbf{AC_1^*}$ *and* **Ex** *are equivalent over* **IST.**

Proof. Assuming $\mathbf{AC_1^*}$, take $\varphi(x,y) \equiv \alpha(y)$ in its antecedent. This yields an $f \in \mathrm{Fun}(1)$ for which $\forall y\alpha(y) \Rightarrow \alpha(f0)$, giving $\exists y[\exists y\alpha(y) \Rightarrow \alpha(y)]$, i.e., **Ex.**

Conversely, define $\alpha(y) \equiv \varphi(0,y)$. Then, assuming **Ex**, there is b for which $\exists y\alpha(y) \Rightarrow \alpha(b)$, whence $\forall x{\in}1\exists y\varphi(x,y) \Rightarrow \forall x{\in}1\varphi(x,b)$.

Defining $f \in \mathrm{Fun}(1)$ by $f = \{\langle 0,b\rangle\}$ gives $\forall x{\in}1\exists y\varphi(x,y) \Rightarrow \forall x{\in}1\varphi(x,fx)$, and $\mathbf{AC}_1^*$ follows. ■

Further, while $\mathbf{DAC}_1$ is easily seen to be provable in **IST**, we have

- $\mathbf{DAC}_1^*$ *and* **Un** *are equivalent over* **IST.**

Proof. Given α, Define $\varphi(x,y) \equiv \alpha(y)$. Then, for $f \in \mathrm{Fun}(1)$, $\exists x{\in}1\varphi(x,fx) \Leftrightarrow \alpha(f0)$ and $\exists x{\in}1\forall y\varphi(x,y) \Leftrightarrow \forall y\alpha(y)$. $\mathbf{DAC}_1^*$ then gives

$$\exists f{\in}\mathrm{Fun}(1)[\alpha(f0) \Rightarrow \forall y\alpha(y)],$$

from which **Un** follows easily.

Conversely, given φ, define $\alpha(y) \equiv \varphi(0,y)$. Then from **Un** we infer that there exists b for which $\alpha(b) \Rightarrow \forall y\alpha(y)$, i.e. $\varphi(0,b) \Rightarrow \forall y\varphi(0,y)$. Defining $f \in \mathrm{Fun}(1)$ by $f = \{\langle 0,b\rangle\}$ then gives $\varphi(0,f0) \Rightarrow \exists x{\in}1\forall y\varphi(x,y)$, whence $\exists x{\in}1\varphi(x,fx) \Rightarrow \exists x{\in}1\forall y\varphi(x,y)$, and **Un** follows. ■

Next, while $\mathbf{AC}_2$ is easily proved in **IST**, by contrast we have

- $\mathbf{DAC}_2$ *and* **Dis** *are equivalent over* **IST**.

Proof. The antecedent of $\mathbf{DAC}_2$ is equivalent to the assertion

$$\forall f{\in}\mathrm{Fun}(2)[\varphi(0, f0) \vee \varphi(1, f1)],$$

which, in view of the natural correlation between members of Fun (2) and ordered pairs, is equivalent to the assertion

$$\forall y \forall y'[\varphi(0, y) \vee \varphi(1, y')].$$

The consequent of $\mathbf{DAC}_2$ is equivalent to the assertion

$$\forall y{\in}Y\varphi(0,y) \vee \forall y'{\in}Y\varphi(1,y')$$

So $\mathbf{DAC}_2$ itself is equivalent to

$$\forall y\forall y'[\varphi(0,y) \vee \varphi(1,y')] \Rightarrow \forall y\varphi(0,y) \vee \forall y'\varphi(1,y').$$

But this is obviously equivalent to the scheme

$$\forall y \forall y'[\alpha(y) \vee \beta(y')] \Rightarrow \forall y \alpha(y) \vee \forall y' \beta(y'),$$

where y does not occur free in β, nor y' in α. And this last is easily seen to be equivalent to **Dis**. ■

Now consider $\mathbf{DAC}_2^*$. This is quickly seen to be equivalent to the assertion

$$\exists z \exists z'[\varphi(0,z) \vee \varphi(1,z') \Rightarrow \forall y \varphi(0,y) \vee \forall y' \varphi(1,y'),$$

i.e. to the assertion, for arbitrary $\alpha(x)$, $\beta(x)$, that

$$\exists z \exists z'[\alpha(z) \vee \beta(z') \Rightarrow \forall y \alpha(y) \vee \forall y' \beta(y')].$$

This is in turn equivalent to the assertion, for any sentence α,

(*) $$\exists y[\alpha \vee \beta(y) \Rightarrow \alpha \vee \forall y \beta(y)] .$$

Now (*) obviously entails **Un.** Conversely, given **Un**, there is b for which $\beta(b) \Rightarrow \forall y \beta(y)$. Hence $\alpha \vee \beta(b) \Rightarrow \alpha \vee \forall y \beta(y)$, whence (*). So we have shown that

- *Over* **IST**, $\mathbf{DAC}_2^*$ *is equivalent to* **Un**, *and hence also to* $\mathbf{DAC}_1^*$.

In order to provide choice schemes equivalent to **Lin** and **Stone** we introduce

$\mathbf{ac}_X^*$ $\quad \exists f \in 2^X [\forall x \in X \exists y \in 2\, \varphi(x,y) \Rightarrow \forall x \in X\, \varphi(x,fx)]$

$\mathbf{wac}_X^*$ $\quad \exists f \in 2^X [\forall x \in X \exists y \in 2\, \varphi(x,y) \Rightarrow \forall x \in X\, \varphi(x,fx)]$ provided $\vdash_{\mathbf{IST}}$ $\forall x[\varphi(x,0) \Rightarrow \neg\varphi(x,1)]$

Clearly $\mathbf{ac}_X^*$ is equivalent to

$$\exists f \in 2^X [\forall x \in X[\varphi(x,0) \vee \varphi(x,1)] \Rightarrow \forall x \in X\, \varphi(x,fx)]$$

and similarly for $\mathbf{wac}_X^*$.

Then

- *Over* **IST**, $\mathbf{ac}_1^*$ *and* $\mathbf{wac}_1^*$ *are equivalent, respectively, to* **Lin** *and* **Stone.**

Proof. Let α and β be sentences, and define $\varphi(x,y) \equiv x = 0 \wedge [(y = 0 \wedge \alpha) \vee (y = 1 \wedge \beta)]$. Then $\alpha \Leftrightarrow \varphi(0,0)$ and $\beta \Leftrightarrow \varphi(0,1)$, and so $\forall x \in 1[\varphi(x,0) \vee \varphi(x,1)] \Leftrightarrow \varphi(0,0) \vee \varphi(0,1) \Leftrightarrow \alpha \vee \beta$. Therefore

$$\begin{aligned} \exists f \in 2^1 [\forall x \in 1[\varphi(x,0) \vee \varphi(x,1)] & \Rightarrow \forall x \in 1\ \varphi(x,fx)] \\ & \Leftrightarrow \exists f \in 2^1[\alpha \vee \beta \Rightarrow \varphi(0,f0)] \\ & \Leftrightarrow [\alpha \vee \beta \Rightarrow \varphi(0,0)] \vee [\alpha \vee \beta \Rightarrow \varphi(0,1)] \\ & \Leftrightarrow [\alpha \vee \beta \Rightarrow \alpha] \vee [\alpha \vee \beta \Rightarrow \beta] \\ & \Leftrightarrow [\beta \Rightarrow \alpha \vee \alpha \Rightarrow \beta]. \end{aligned}$$

This yields $\mathbf{ac}_1^* \Rightarrow$ **Lin.** For the converse, define $\alpha \equiv \varphi(0,0)$ and $\beta \equiv \varphi(0,1)$ and reverse the argument.

To establish the second stated equivalence, notice that, when $\varphi(x,y)$ is defined as above, but with β replaced by $\neg\alpha$, it satisfies the provisions imposed in $\mathbf{wac}_1^*$. As above, that principle gives $(\neg\alpha \Rightarrow \alpha) \vee (\alpha \Rightarrow \neg\alpha)$, whence $\neg\alpha \vee \neg\neg\alpha$. So **Stone** follows from $\mathbf{wac}_1^*$. Conversely, suppose that φ meets the condition imposed in $\mathbf{wac}_1^*$. Then from $\varphi(0,0) \Rightarrow \neg\varphi(0,1)$ we deduce $\neg\neg\varphi(0,0) \Rightarrow \neg\varphi(0,1)$; now, assuming **Stone**, we have $\neg\varphi(0,0) \vee \neg\neg\varphi(0,0)$, whence $\neg\varphi(0,0) \vee \neg\varphi(0,1)$. Since $\neg\varphi(0,0) \Rightarrow [\varphi(0,0) \Rightarrow \varphi(0,1)]$ and $\neg\varphi(0,1) \Rightarrow [\varphi(0,1) \Rightarrow \varphi(0,0)]$ we deduce $[\varphi(0,0) \Rightarrow \varphi(0,1)] \vee [\varphi(0,1) \Rightarrow \varphi(0,0)]$. From the argument above it now follows that $\exists f \in 2^1 [\forall x \in 1[\varphi(x,0) \vee \varphi(x,1)] \Rightarrow \forall x \in 1\ \varphi(x,fx)]$. Accordingly $\mathbf{wac}_1^*$ is a consequence of **Stone**.

AC AND HILBERT'S ε-CALCULUS

Hilbert's investigations into the foundations of mathematics in the 1920s had led him to regard **AC** as an indispensable principle which he believed would prove useful in his defense of classical mathematics against the attacks of the intuitionists.[1] In his foundational framework **AC** took the form of a postulate he called *the logical ε-axiom.*

To formulate his postulate, Hilbert introduced, for each formula $\varphi(x)$[2], a term $\varepsilon_x\varphi$ which, intuitively, is intended to name an indeterminate object satisfying $\varphi(x)$. Then Hilbert's ε-axiom reads:

(ε) $$\varphi(x) \Rightarrow \varphi(\varepsilon_x\varphi).$$

In any of the usual logical systems this is equivalent to

(ε*) $$\exists x\varphi(x) \Rightarrow \varphi(\varepsilon_x\varphi)[3].$$

Accordingly all that is known about $\varepsilon_x\varphi$ is that, if anything satisfies φ, it does[4].

It can now be seen how **AC** emerges from all this. Since φ may contain free variables other than x, the identity of $\varepsilon_x\varphi$ depends, in general, on the values assigned to these variables. So $\varepsilon_x\varphi$ may be regarded as the result of having chosen, for each assignment of values to these other variables, a value of x so that $\varphi(x)$ is satisfied. That is, $\varepsilon_x\varphi$ may be construed as a choice function, and the ε-axiom accordingly seen as a version of **AC.**

---

[1] It is therefore somewhat ironic that **AC** - at least in its "logical" form - is affirmable intuitionistically: see the following section and chapter.

[2] The formula φ may have more than one free variable.

[3] It should be noted that in introducing ε-terms Hilbert's principal purpose was to provide a concrete way of *defining* the existential quantifier (which he regarded as a "transfinite" notion). Thus, in his system, $\exists x\varphi(x)$ was simply another way of writing $\varphi(\varepsilon_x\varphi)$. This is precisely the strategy adopted by Bourbaki in their Elements de Mathematique (except, as has already been observed, they use "τ" in place of "ε").

[4] David Devidi has had the happy inspiration of calling $\varepsilon_x\varphi$ "the thing most likely to be φ."

In general, an ε-*calculus* T_ε is obtained by starting with a first-order theory T, augmenting its language $\mathscr{L}$ with epsilon terms, and adjoining to T as an axiom scheme the formulas **(ε*).** It is known that when T is any classical first-order theory, T_ε is *conservative* over T, that is, each assertion of $\mathscr{L}$ demonstrable in T_ε is also demonstrable in T: the move from T to T_ε does not enlarge the body of demonstrable assertions in T[1] . But for *intuitionistic* predicate logic the situation is decidedly otherwise.

In fact it can be shown that, if T is taken to be intuitionistic predicate calculus IPC, then a number of assertions undemonstrable within I, for instance **Ex** and **Lin** above, become provable within IPC_ε . On the other hand, **SLEM** is *not* derivable in IPC_ε[2]. This is related to the fact (remarked on above) that in deriving **LEM** from **AC** one requires the Principle of Extensionality for Functions. The analogous principle within the ε-calculus is the *Principle of Extensionality for ε-terms*:

Ext$_\varepsilon$ $\qquad \forall x[\varphi(x) \Leftrightarrow \psi(x)] \Rightarrow \varepsilon_x\varphi = \varepsilon_x\psi.$

An argument similar to the derivation of **LEM** from **AC** given above yields **SLEM** from **Ext$_\varepsilon$** within (a very weak extension of) I_ε . In brief, the argument runs as follows. We augment IPC_ε by **Ext$_\varepsilon$** and the sentence $0 \neq 1$, and argue informally within the resulting theory[3]. Thus let σ be any sentence and let $\varphi(x)$, $\psi(x)$ be the formulas $(x = 0) \vee \sigma$, $(x = 1) \vee \sigma$ respectively. Then clearly $\exists x\varphi$ and $\exists x\psi$, whence $\varphi(\varepsilon_x\varphi)$ and $\psi(\varepsilon_x\psi)$. This means that $(\varepsilon_x\varphi = 0) \vee \sigma$ and $(\varepsilon_x\psi = 1) \vee \sigma$, whence $(\varepsilon_x\varphi = 0 \ \wedge \varepsilon_x\psi = 1) \ \vee \sigma$ from which it follows that (*) $\varepsilon_x\varphi \neq \varepsilon_x\psi \vee \sigma$. But clearly $\sigma \Rightarrow \forall x[\varphi(x) \Leftrightarrow \psi(x)]$, so

---

[1] This is the *second* ε-*theorem* of Hilbert-Bernays. See, e.g. Kneebone [1963] or Leisenring [1969].

[2] Bell [1993].

[3] In some of the arguments below we shall employ this augmented theory without comment.

we deduce from (Ext) that $\sigma \Rightarrow \varepsilon_x\varphi = \varepsilon_x\psi$. Therefore $\varepsilon_x\varphi \neq \varepsilon_x\psi \Rightarrow \neg\sigma$ and it now follows from (*) that $\neg\sigma \vee \sigma$. This gives **SLEM.**

The use of $\mathbf{Ext}_\varepsilon$ can be avoided in deriving **SLEM** in IPC_ε if one takes **(ε*)** in the (classically equivalent) form

(ε)** $$\varphi(\varepsilon_x\varphi) \vee \forall x \neg\varphi(x).$$

This may be read: "either $\varepsilon_x\varphi$ satisfies φ or nothing does". From this we can intuitionistically derive **SLEM** as follows:

Given a sentence σ, define $\varphi(x)$ to be the formula

$$(x = 0 \wedge \sigma) \vee (x = 1 \wedge \neg\sigma).$$

Then from (ε**) we get

$$[(\varepsilon_x\varphi = 0 \wedge \sigma) \vee ([(\varepsilon_x\varphi = 1 \wedge \neg\sigma)] \vee \forall x \neg[(x = 0 \wedge \sigma) \vee (x = 1 \wedge \neg\sigma)],$$

which implies

$$[\sigma \vee \neg\sigma) \vee [\forall x \neg(x = 0 \wedge \sigma) \wedge \forall x \neg(x = 1 \wedge \neg\sigma)],$$

whence

$$[\sigma \vee \neg\sigma) \vee [\neg\, \sigma \wedge \neg\neg\sigma],$$

winding up with $\sigma \vee \neg\sigma$.

The use of $\mathbf{Ext}_\varepsilon$ can be also avoided in deriving **SLEM** in IPC_ε if one employs *relative* ε-*terms,* that is, allows ε to act on *pairs* of formulas, each with a *single* free variable. Here, for each pair of formulas $\varphi(x)$, $\psi(x)$ we introduce the "relativized" ε-term $\varepsilon_x\varphi/\psi$ and the "relativized" ε-axioms

(1) $\exists x\, \psi(x) \Rightarrow \psi(\varepsilon_x\varphi/\psi)$ (2) $\exists x\, [\varphi(x) \wedge \psi(x)] \Rightarrow \varphi(\varepsilon_x\varphi/\psi)$.

That is, $\varepsilon_x\varphi/\psi$ may be thought of as an individual that satisfies ψ if anything does, and which in addition satisfies φ if anything satisfies both φ and ψ. Notice that the usual ε–term $\varepsilon_x\varphi$ is then $\varepsilon_x\varphi/(x = x)$. In the classical ε-calculus $\varepsilon_x\varphi/\psi$ may be defined by taking

$$\varepsilon_x\varphi/\psi \;=\; \varepsilon_y[[y = \varepsilon_x(\varphi \wedge \psi) \wedge \exists x\, (\varphi \wedge \psi)] \vee [y = \varepsilon_x\psi \wedge \neg\exists x\, (\varphi \wedge \psi)]].$$

But the relativized ε-axioms are not derivable in IPC_ε since they can be shown to imply **SLEM**. To see this, given a sentence σ define

$$\varphi(x) \equiv x = 1 \qquad \psi(x) \equiv x = 0 \vee \sigma.$$

Write a for $\varepsilon_x\varphi/\psi$. Then we certainly have $\exists x\psi(x)$, so (1) gives $\psi(a)$, i.e.

(3) $$a = 0 \vee \sigma$$

Also $\exists x\,(\varphi \wedge \psi) \Leftrightarrow \sigma$, so (2) gives $\sigma \Rightarrow \varphi(a)$, i.e.

$$\sigma \Rightarrow a = 1,$$

whence

$$a \neq 1 \Rightarrow \neg\sigma,$$

so that

$$a = 0 \Rightarrow \neg\sigma.$$

And the conjunction of this with (3) gives $\sigma \vee \neg\sigma$, as claimed.

The relationship between the ε-operator and *set theory* may be briefly described as follows. If one takes a classical system of set theory such as **ZF,** augments the language with ε-terms and simply adds the scheme **(ε*)** to the axioms of **ZF**, then one obtains a theory $\mathbf{ZF}_\varepsilon$ which is conservative over **ZF.** On the other hand, if ε-terms are *permitted to appear in the Axiom Schemes of Separation and Replacement,* then a theory $\mathbf{ZF}_\varepsilon$* is obtained *in which* **AC** *is derivable.* For under these conditions we have, for any formula $\varphi(x,y)$,

(*) $$\forall x{\in}X\, \exists y\, \varphi(x,y) \Rightarrow \forall x{\in}X\, \varphi(x, \varepsilon_y\varphi(x,y))\ .$$

Let $t \in Fun(X)$ be the map $x \mapsto \varepsilon_y\varphi(x,y)$: the Axiom of Replacement applied to the term $\varepsilon_y\varphi(x,y)$ guarantees that t is a function on X. Then, from (*)

$$\forall x{\in}X\, \exists y\, \varphi(x,y) \Rightarrow \forall x{\in}X\, \varphi(x, tx)\,,$$

so that

$$\exists f{\in}Fun\,(X)\, [\forall x{\in}X\, \exists y\, \varphi(x,y) \Rightarrow \forall x{\in}X\, \varphi(x, fx)],$$

i.e. $\mathbf{AC}_X^*$.

In the case of intuitionistic Zermelo-Fraenkel set theory **IZF**[1], the situation differs somewhat from its classical analogue. To begin with, augmenting the language with ε-terms and simply adding the scheme **(ε*)** to the axioms of **IZF** without allowing ε-terms to appear in the Axiom Schemes of Separation or Replacement, yields a theory $\mathbf{IZF}_\varepsilon$ which is not conservative over **IZF**, since, as we have seen, in any similarly augmented intuitionistic theory, one can prove such formerly underivable logical "choice" principles as **Ex**. On the other hand, just as in the classical case, the move fails to produce **AC.** The analogy with the classical case is strengthened when one considers what happens when ε-terms are allowed to appear in the Axiom Schemes of Separation and Replacement, generating the corresponding theory $\mathbf{IZF}_\varepsilon^*$. In $\mathbf{IZF}_\varepsilon^*$, using precisely the same argument as above, **AC** becomes derivable, and so therefore **LEM**. But **IZF** augmented by **LEM** is just classical **ZF.** So, as with **AC,** the "logical" effect of adding the ε-axiom (in the appropriate way) to **IZF** is to transform it into its classical analogue.

Finally, we point out that the ε-operator was not in fact the first device introduced by Hilbert to justify the use of classical reasoning in mathematics. For in 1923 he introduced what amounts to a dual form of the ε-operator, the *τ-operator,* which was governed by a principle he called the *Transfinite Axiom*[2]

Trans $\qquad \varphi(\tau_x\varphi) \Rightarrow \varphi(x).$

In any of the usual logical systems this is equivalent to

(*) $\qquad \varphi(\tau_x\varphi) \Rightarrow \forall x\alpha(x).$

---

[1] This is the intuitionistic theory whose axioms are those of **ZF**, but in which the Axiom of Regularity (which, as it happens, implies **LEM**) has been replaced by the principle of ∈-induction.

[2] See section 4.8 of Moore [1982].

Accordingly all that is known about $\tau_x\varphi$ is that, if it satisfies φ, anything does[1].

It is an easy matter to derive **Un** above from the τ-scheme when τ is merely allowed to act on formulas with at most one free variable. When τ's action is extended to formulas with two free variables, the τ-scheme applied in **IST** yields the full dual axiom of choice $\forall X$ $\mathbf{DAC}^*_X$. For under these conditions we have, for any formula $\varphi(x,y)$,

(*) $$\forall x \in X[\varphi(x,\tau_y\varphi(x,y)) \Rightarrow \forall y\varphi(x,y)]$$

Let $t \in$ Fun(X) be the map $x \mapsto \tau_y\varphi(x,y)$. Assuming that $\forall f \in Y^X \exists x \in X \varphi(x, fx)$, let $a \in X$ satisfy $\varphi(a,ta)$. We deduce from (*) that $\forall y \in Y\varphi(a,y)$, whence $\exists x \in X \forall y \in Y \varphi(x, y)$. The dual axiom of choice follows.

Accordingly the τ-operator bears the same relationship to the Dual Axiom of Choice as does the ε-operator to the Axiom of Choice itself.

AC AND THE LAW OF EXCLUDED MIDDLE IN WEAK INTUITIONISTIC SET THEORIES

As mentioned in the Introduction, a new twist has arisen in the story of **AC** as the result of certain new developments in systems of constructive mathematics, in particular Constructive Type Theory.[2] This twist can be most easily described by considering the principle we have labelled **AC3**, viz.,

for any relation R between sets A, B,

$$\forall x \in A \exists y \in B\ R(x,y) \Rightarrow \exists f\colon A \to B\ \forall x \in A\ R(x, fx).$$

[1] To enlarge on Devidi's suggestion, $\tau_x\varphi$ is "the thing least likely to be φ".

[2] We defer until Chapter VII formal discussion of Constructive Type Theory and the role of **AC** therein.

Now under the strictly constructive interpretation of quantifiers implicit in intuitionistic mathematics, and later given explicit form in Constructive Type Theory, the assertability of an alternation of quantifiers $\forall x \exists y R(x,y)$ means *precisely* that one is given a function f for which $R(x,fx)$ holds for all x. In the words of Bishop [1967], *a choice function exists in constructive mathematics because a choice is implied by the very meaning of existence.* Thus, for example, the antecedent $\forall x \in A \exists y \in B\ R(x, y)$ of **AC3,** given a constructive construal, just *means* that we have a procedure which, applied to each $x \in A$, yields a $y \in B$ for which $R(x, y)$. But this is precisely what is expressed by the consequent $\exists f\colon A \to B\ \forall x \in A\ R(x, fx)$of **AC3**.

It follows that **AC3** is actually *derivable* in such constructive settings. On the other hand this is decidedly not the case for **LEM**. This incongruity has been the subject of a number of recent investigations[1]. What has emerged is that for the derivation of **LEM** from **AC** to go through it is sufficient that sets (in particular power sets), or functions, have a degree of extensionality which is, so to speak, built into the usual set theories but is incompatible with Constructive Type Theory. Another condition, independent of extensionality, ensuring that the derivation goes through is that any equivalence relation determines a quotient set . **LEM** can also be shown to follow from a suitably extensionalized version of **AC**. The arguments establishing these intriguing results were origuinally formulated within Constructive Type Theory. In this section we shall derive analogous results within a comparatively straightforward *set-theoretic* framework[2]. The core principles of

[1] See for example Maietti [1999], Maietti and Valentini [1999], Martin-Löf [2006], and Valentini [2002].

[2] Bell [2008].

this framework form a theory - *weak set theory* **WST** - which is based on intuitionistic logic, lacks the axiom of extensionality[1], and supports only minimal set-theoretic constructions[2]. **WST** is, like Constructive Type Theory, too weak to allow the derivation of **LEM** from **AC.** But we shall see that, as with Constructive Type Theory, beefing up **WST** with extensionality principles or quotient sets enables the derivation to be carried out.

Let $\mathscr{L}^{<,>}$ be the first-order language $\mathscr{L}$ of (intuitionistic) set theory augmented with a binary operation symbol $\langle\, ,\, \rangle$ permitting the formation of ordered pairs. At certain points various additional predicates and operation symbols will be introduced into $\mathscr{L}^{<,>}$. The restricted quantifiers $\exists x{\in}a$ and $\forall x{\in}a$ are defined as usual, that is, as $\exists x(x \in a \wedge \ldots)$ and $\forall x(x \in a \rightarrow \ldots)$ respectively. A formula is *restricted* if it contains only restricted quantifiers.

Weak set theory **WST** is the theory in $\mathscr{L}^{<,>}$ with the following basic axioms (in which the free variables are understood to be universally quantified, and similarly below):

Unordered Pair $\exists u \forall x[x \in u \Leftrightarrow x = a \vee x = b]$

Ordered Pair $\langle a,b\rangle = \langle c,d\rangle \Leftrightarrow a = c \wedge b = d$

Binary Union $\exists u \forall x[x \in u \Leftrightarrow x \in a \vee x \in b]$

Cartesian Product $\exists u \forall x[x \in u \Leftrightarrow \exists y \in a \exists z \in b\ (x = \langle y,z\rangle)]$

Restricted Separation $\exists u \forall x[x \in u \Leftrightarrow x \in a \wedge \varphi(x)]$

where in this last axiom φ is any restricted formula with at most the variable x free.

Rudimentary set theory **RST** is obtained from **WST** by confining Restricted Separation to atomic and negated atomic formulas.

---

[1] Set theories (with classical logic) lacking the Axiom of Extensionality seem first to have been extensively studied in Gandy [1956, 1959], Scott [1966].

[2] **WST** may be considered a fragment both of (intuitionistic) Δ_0-Zermelo set theory and Aczel's constructive set theory (Aczel and Rathjen 2001).

We introduce into $\mathcal{L}^{<,>}$ new predicates and operation symbols as indicated below and adjoin to **WST** by the following "definitional" axioms:

$a \subseteq b \Leftrightarrow \forall x[x \in a \Rightarrow x \in b]$ $\quad a \approx b \Leftrightarrow \forall x[x \in a \Leftrightarrow x \in b]$ $\quad Ext(a) \Leftrightarrow \forall x \in a \forall y \in a[x \approx y \Rightarrow x = y]$

$x \in a \cup b \Leftrightarrow x \in a \vee x \in b$ $\quad x \in \{a,b\} \Leftrightarrow x = a \vee x = b$ $\quad \{a\} = \{a,a\}$ $\quad x\ r\ y \Leftrightarrow \langle x,y \rangle \in r$

$detach(b,a) \Leftrightarrow b \subseteq a \wedge \forall x \in a[x \in b \vee x \notin b]$

$y \in \{x \in a : \varphi(x)\} \Leftrightarrow y \in a \wedge \varphi(y)$

$\neg x \in 0 \quad 1 = \{0\} \quad 2 = \{0,1\}$

$x \in a \times b \Leftrightarrow \exists u \in a \exists v \in b(x = \langle u,v \rangle)$ $\quad x \in a + b \Leftrightarrow \exists u \in a \exists v \in b[x = \langle u,0 \rangle \vee x = \langle v,1 \rangle]$

$f : a \to b \Leftrightarrow f \subseteq a \times b \wedge \forall x \in a \exists y \in b(x\ f\ y) \wedge \forall x \forall y \forall z[(x\ f\ y \wedge x\ f\ z) \Rightarrow y = z]$

$Fun(f) \Leftrightarrow \exists a \exists b(f : a \to b)$ $\quad f : a \to b \wedge x \in a \Rightarrow x\ f\ f(x)$

$f : a \to b \wedge g : b \to c \Rightarrow g \circ f : a \to c \wedge \forall x \in a[(g \circ f)(x) = g(f(x))]$

$f : a \twoheadrightarrow b \Leftrightarrow f : a \to b \wedge \forall y \in b \exists x \in a[y = f(x)]$

$\pi_1 : a + b \to a \cup b \wedge \forall x \in a[\pi_1(\langle x,0 \rangle) = x] \wedge \forall y \in b[\pi_1(\langle y,1 \rangle) = y]$

$\pi_2 : a + b \to 2 \wedge \forall x \in a[\pi_2(\langle x,0 \rangle) = 0] \wedge \forall y \in b[\pi_2(\langle y,1 \rangle) = 1]$

$Eq(s,a) \Leftrightarrow s \subseteq a \times a \wedge \forall x \in a(x\ s\ x) \wedge \forall x \in a \forall y \in a(x\ s\ y \Rightarrow y\ s\ x) \wedge$
$\forall x \in a \forall y \in a \forall z \in a[(x\ s\ y \wedge y\ s\ z) \Rightarrow x\ s\ z]$

$Comp(r,s) \Leftrightarrow \forall x \forall x' \forall y[(x\ s\ x' \wedge x'\ r\ y) \Rightarrow x\ r\ y]$

$Comp(r) \Leftrightarrow \forall x \forall x' \forall y[(x \approx x' \wedge x'\ r\ y) \Rightarrow x\ r\ y]$

$Extn(f,s) \Leftrightarrow Fun(f) \wedge \forall x \forall x'[x\ s\ x' \wedge \exists y \exists y'(x\ f\ y \wedge x'\ f\ y') \Rightarrow f(x) = f(x')]$

$Ex(f) \Leftrightarrow Fun(f) \wedge \forall x \forall x'[x \approx x' \wedge \exists y \exists y'(x\ f\ y \wedge x'\ f\ y') \Rightarrow f(x) = f(x')]$

Most of these definitions are standard. The functions π_1 and π_2 are projections of ordered pairs onto their 1st and 2nd coordinates respectively: clearly, for u, $v \in a + b$ we have

(proj) $\qquad u = v \Leftrightarrow [\pi_1(u) = \pi_1(v) \wedge \pi_2(u) = \pi_2(v)]$.

The relation $\approx$ is that of *extensional equality.* Ext(a) expresses the *extensionality* of the members of the set a. $Eq(s,a)$ asserts that s is an equivalence relation on a. If r is a relation between a and b, and s an relation on a, *Comp*(r,s) expresses the *compatibility* of r with s, and *Comp*(r) the compatibility of r with extensional equality. If $f: a \to b$, and s is an equivalence relation on a, *Etxn*(f,s) expresses the idea that f treats the relation s as if it were the identity relation: we shall then say that f is *s-extensional*. $Ex(f)$ asserts that f is *extensional* in the sense of treating extensional equality as if it were identity. Finally $Detach(b, a)$ says that b is a *detachable* subset of a, in other words, that b has a genuine complement in a.

We formulate the following axioms additional to those of **WST**:

Extensionality $a \approx b \Rightarrow a = b$

Detachability $\mathrm{b} \subseteq \mathrm{a} \Rightarrow \mathrm{detach(b, a)}$

This is essentially Excluded Middle for formulas of the form $x \in b$.

Extpow(1) $\exists u[\forall x(x \in u \Leftrightarrow x \subseteq 1) \wedge Ext(u)$

This asserts that 1 has an extensional power set. In **WST** + **Extpow(1),** we introduce the new term Ω and adjoin the "definitional" axiom

(Ω) $\forall x(x \in \Omega \Leftrightarrow x \subseteq 1) \wedge Ext(\Omega)$.

Our next axiom is

Extdoub(2) $a \subseteq 2 \wedge b \subseteq 2 \Rightarrow Ext(\{a, b\})$

This asserts that all doubletons composed of subsets of 2 are extensional.

Next, three versions of the Axiom of Choice:

Axiom of Choice AC[1]

$$r \subseteq a \times b \wedge \forall x \in a \exists y \in b(x\ r\ y) \Rightarrow \exists f : a \to b \forall x \in a(x\ r\ f(x))$$

Universal Extensional Axiom of Choice UEAC

$$Eq(s, a) \wedge r \subseteq a \times b \wedge Comp(r, s) \wedge \forall x \in a \exists y \in b(x\ r\ y) \Rightarrow$$
$$\exists f : a \to b[Extn(f, s) \wedge \forall x \in a(x\ r\ f(x))]$$

[1] This is essentially what in Chapter 1 we called **AC3**. For simplicity we drop the "**3**" here.

Extensional Axiom of Choice EAC

$$r \subseteq a \times b \wedge Comp(r) \wedge \forall x \in a \exists y \in b(x\ r\ y) \Rightarrow$$
$$\exists f : a \to b[Ex(f) \wedge \forall x \in a(x\ r\ f(x))]$$

AC asserts, as usual, that a choice function always exists under the appropriate conditions on a given relation *r*. **UEAC** further asserts that, in the presence of an equivalence relation *s* with which *r* is compatible, the choice function can be taken to be *s*-extensional. **AC** can be seen to be the special case of **UEAC** in which *s* is taken to be the identity relation: for this reason **AC** is sometimes known as the *Intensional* Axiom of Choice. Finally **EAC** is the special case of **UEAC** in which the equivalence relation is that of extensional equality.

Our next axiom is

Quotients

$$Eq(s,a) \Rightarrow \exists u \exists f[f : a \twoheadrightarrow u \wedge \forall x \in a \forall y \in a[f(x) = f(y) \Leftrightarrow x\ s\ y]]$$

This axiom asserts that each equivalence relation determines a quotient set. In **WST + Quotients,** we introduce operation symbols $\cdot/_\cdot$, $[\cdot]_\cdot$ and adjoin the "definitional" axiom

(Q)

$$Eq(s,a) \Rightarrow [\forall x \in a([x]_s \in a/s) \wedge \forall u \in a/_s \exists x \in a(u = [x]_s)$$
$$\wedge \forall x \in a \forall y \in a[[x]_s = [y]_s \Leftrightarrow x\ s\ y]]$$

Here $a/_s$ is the *quotient* of *a* by *s* and, for $x \in a$, $[x]_s$ is the *image* of *x* in a/s.

Reminding the reader that our background logic is intuitionistic, we finally introduce the following logical schemes:

Restricted Excluded Middle for Sentences REMS

$\sigma \vee \neg\sigma$ for any restricted sentence σ

Restricted Excluded Middle REM

$\forall x \in a[\varphi(x) \vee \neg\varphi(x)]$ for any restricted formula φ with at most the variable *x* free

Now let **WSTC** = **WST** + **AC**, **WSTEC** = **WST** + **EAC**, **RSTC** = **RST** + **AC** , and **WSTQ** = **WST** + **Quotients.** We are going to prove the following:

Theorem 1[1]**. REMS** *is derivable in* **(a) WSTC + Extpow(1), (b) WSTC + Extdoub(2),** *and* **(c) WSTEC.**

Theorem 2. (i) Detachability *is derivable* in **RSTC + Quotients. (ii) REM** *is derivable in* **WSTC + Quotients**

Theorem 3. AC ⇔ UEAC *is derivable in* **WSTQ**.

Thus, while in the absence of extensional power sets and extensional doubletons, the Intensional Axiom of Choice does not yield Excluded Middle, it does so in the presence of either of the former. Moreover, the Extensional Axiom of Choice always entails the Excluded Middle. And finally, when quotients are present the Intensional Axiom of Choice is no weaker than its Universal Extensional version.

Proof of Theorem 1.

(a) We argue in **WSTC** + **Extpow(1).** Recalling **(Ω)** above, we define[2]

$$a = \{\langle u,v\rangle \in \Omega\times\Omega : 0 \in u \cup v\}$$

Then clearly

$$\langle u,v\rangle \in a \Rightarrow \exists x \in 2[(x=0 \Rightarrow 0 \in u) \wedge (x=1 \Rightarrow 0 \in v)].$$

So **AC** gives $f\colon a \to 2$ such that , for $\langle u,v\rangle \in a$

(1) $$f(\langle u,v\rangle) = 0 \to 0 \in u$$

(2) $$f(\langle u,v\rangle) = 1 \Rightarrow 0 \in v.$$

Also, for $\langle u,v\rangle \in a$, we have

[1] Theorems 1 and 2 may be seen as precise versions of the derivations of **AC** from the various principles introduced at the beginning of this chapter.

[2] Here the expression on the right hand side is an abbreviation for $\{z \in \Omega\times\Omega : 0 \in \pi_1(z) \cup \pi_1(z)\}$. Similar abbreviations will be used in the sequel.

(3) $$f(\langle u, v\rangle) = 0 \vee f(\langle u, v\rangle) = 1.$$

Now for *arbitrary* $u \in \Omega$ we have $\langle u, \{0\}\rangle \in a$ and $\langle \{0\}, u\rangle \in a$. Substituting $\{0\}$ for u in (3) and using (1) gives

$$0 \in u \vee f(\langle u, \{0\}\rangle) = 1.$$

Similarly, substituting u for v and $\{0\}$ for u in (3) and using (2) gives

$$f(\langle \{0\}, u\rangle) = 0 \vee 0 \in u.$$

Conjoining these last two assertions and applying the distributive law yields

(4) $$0 \in u \vee [f(\langle \{0\}, u\rangle) = 0 \wedge f(\langle u, \{0\}\rangle) = 1].$$

Writing $\varphi(u)$ for the second disjunct in (4), the latter then becomes

(5) $$0 \in u \vee \varphi(u).$$

From $u \in \Omega$ we deduce

$$0 \in u \Rightarrow u \approx \{0\},$$

and so, since (again recalling (**Ω**) above) $Ext(\Omega)$,

$$0 \in u \Rightarrow u = \{0\}.$$

Hence

$$\begin{aligned}[0 \in u \wedge \varphi(u)] &\Rightarrow [u = \{0\} \wedge \varphi(u)] \\ &\Rightarrow \varphi(\{0\}) \\ &\Rightarrow 0 = 1.\end{aligned}$$

Since clearly $0 \neq 1$, we conclude that

$$\varphi(u) \Rightarrow \neg 0 \in u$$

and (5) then yields

(6) $$0 \in u \vee \neg 0 \in u.$$

This holds for arbitrary $u \in \Omega$. So, given a restricted sentence σ, define $s = \{x \in \{0\}: \sigma\}$. Then $s \in \Omega$ and $0 \in s \Leftrightarrow \sigma$. It now follows from (6) that $\sigma \vee \neg\sigma$, as required.

(b) We argue in **WSTC +Extdoub(2)**. Given a restricted sentence σ, define

$$a = \{x \in 2 : x = 0 \vee \sigma\}, \quad b = \{x \in 2 : x = 1 \vee \sigma\}.$$

Then $a \subseteq 2$ and $b \subseteq 2$ and **Extdoub(2)** gives *Ext{a, b}*. Since $0 \in a$ and $1 \in b$, we have

$$\forall x \in \{a,b\} \exists y \in 2. y \in x,$$

and so **AC** applied to the relation

$$r = \{\langle x,y\rangle \in \{a,b\} \times 2 : y \in x\})$$

yields a function f: $\{a, b\} \rightarrow 2$ for which $\forall x \in \{a,b\}. f(x) \in x$. It follows that $f(a) \in a \wedge f(b) \in b$, so that

$$[f(a) = 0 \vee \sigma] \wedge [f(b) = 1 \vee \sigma].$$

Applying the distributive law, we then get

$$\sigma \vee [f(a) = 0 \wedge f(b) = 1].$$

whence

(1) $$\sigma \vee f(a) \neq f(b).$$

Now clearly $\sigma \Rightarrow a \approx b$, and from this and *Ext*({*a*,*b*}) we deduce $\sigma \Rightarrow a = b$, whence

(2) $$\sigma \Rightarrow f(a) = f(b).$$

It follows that $f(a) \neq f(b) \Rightarrow \neg\sigma$, and we conclude from (1) that $\sigma \vee \neg\sigma$, as required.

(c) Here the argument in **WSTEC** is the same as that given in **(b)** except that in deriving (2) above we invoke **EAC** in place of **Extdoub(2).** To justify this step it suffices to show that *Comp*(*r*), where *r* is the relation defined in the proof of **(b)**. This, however, is clear. ■

Proof of Theorem 2. [1]

[1] The proof of Theorem 2 is an adaptation to a set-theoretical context of the argument in Diaconescu [1975] that, in a topos satisfying **AC**, all subobjects are complemented. By weakening **Quotients** to the assertion **Quotients(1 + 1)** that quotient sets are determined just by equivalence relations on the set 1 + 1, the proof of Theorem 2 shows that **REM** is derivable in the theory **WSTC + Quotients(1 + 1).**

(i) Let us call an *indicator* for a subset b of a any function g: $a \times 2 \to 2$ satisfying

$$\forall x \in a[x \in b \Leftrightarrow g(\langle x,0\rangle) = g(\langle x,1\rangle)].$$

It is easy to show that a subset is detachable if and only if it has an indicator. For if $b \subseteq a$ *is* detachable, then g: $a \times 2 \to 2$ defined by

$$g(\langle x,0\rangle) = g(\langle x,1\rangle) = 0 \qquad \text{if } x \in b$$
$$g(\langle x,0\rangle) = 0 \wedge g(\langle x,1\rangle) = 1 \qquad \text{if } x \notin b$$

is an indicator for b. Conversely, for any function g: $a \times 2 \to 2$, we have $g(\langle x,0\rangle) = g(\langle x,1\rangle) \vee g(\langle x,0\rangle) \neq g(\langle x,1\rangle)$, so if g is an indicator for b, we infer $\forall x \in a[x \in b \vee x \notin b]$, and u is detachable.

Now we show in **RSTC + Quotients** that every subset of a set has an indicator, and is accordingly detachable. For $b \subseteq a$, let s be the binary relation on $a + a$ given by:

$$s = \{\langle\langle x,0\rangle,\langle x,0\rangle\rangle : x \in a\} \cup \{\langle\langle x,1\rangle,\langle x,1\rangle\rangle : x \in a\} \cup$$
$$\{\langle\langle x,0\rangle,\langle x,1\rangle\rangle : x \in b\} \cup \{\langle\langle x,1\rangle,\langle x,0\rangle\rangle : x \in b\}.$$

It is easily checked that $Eq(s, a + a)$. Also, it is clear that, for $z, z' \in a + a$,

(1) $$z\ s\ z' \Rightarrow \pi_1(z) = \pi_1(z')$$

and, for $x \in a$,

(2) $$x \in b \Leftrightarrow \langle x,0\rangle\ s\ \langle x,1\rangle.$$

Invoking axiom (**Q**) above, we introduce the quotient ${(a+a)}/{s}$ of $a + a$ by s and the image $[u]_s$ of an element u of $a + a$ in ${(a+a)}/{s}$ for which we then have

(3) $$\forall z \in {(a+a)}/{s}\ \exists u \in a + a(z = [u]_s)$$

and

(4) $$\forall u \in a + a \forall v \in a + a[[u]_s = [v]_s \Leftrightarrow u\ s\ v].$$

Applying **AC** to (3) yields a function $f : {(a+a)}/{s} \to a + a$ for which

(5) $$\forall z \in {}^{(a+a)}\!/_{s}\; z = [f(z)]_s).$$

Clearly *f is* one-one, that is, we have

(6) $$f(z) = f(z') \Leftrightarrow z = z'.$$

Next, observe that, for $i = 0, 1$, and $x \in a$,

(7) $$\pi_1(f([\langle x,i\rangle]_s)) = x.$$

For from (5) we have $[\langle x,i\rangle]_s = [f([\langle x,i\rangle]_s)]_s$, whence by (4) $\langle x,i\rangle\ s\ f([\langle x,i\rangle]_s)$. Hence by (1) $\pi_1(\langle x,i\rangle) = \pi_1(f([\langle x,i\rangle]_s))$. (7) now follows from this and the fact that $\pi_1(\langle x,i\rangle) = x$.

We have also

(8) $$x \in b \Leftrightarrow f([\langle x,0\rangle]_s) = f([\langle x,1\rangle]_s).$$

For we have

$$\begin{aligned} x \in b &\Leftrightarrow \langle x,0\rangle\ s\ \langle x,1\rangle && \text{using (2)} \\ &\Leftrightarrow [\langle x,0\rangle]_s = [\langle x,1\rangle]_s && \text{using (4)} \\ &\Leftrightarrow f([\langle x,0\rangle]_s) = f([\langle x,1\rangle]_s) && \text{using (6).} \end{aligned}$$

Now define $g: a \times 2 \to 2$ by

$$g(\langle x,i\rangle) = \pi_2(f([\langle x,i\rangle]_s)).$$

We claim that *g* is an indicator for *b*. This can be seen from the following equivalences:

$$\begin{aligned} x \in b &\Leftrightarrow f([\langle x,0\rangle]_s) = f([\langle x,1\rangle]_s) && \text{(by (8))} \\ &\Leftrightarrow \pi_1(f([\langle x,0\rangle]_s)) = \pi_1(f([\langle x,1\rangle]_s)) \\ &\quad \wedge \pi_2(f([\langle x,0\rangle]_s)) = \pi_2(f([\langle x,1\rangle]_s)) \\ &\quad \text{(by } \textbf{(proj)}\text{)} \\ &\Leftrightarrow \pi_2(f([\langle x,0\rangle]_s)) = \pi_2(f([\langle x,1\rangle]_s)) \\ &\quad \text{(using (7))} \\ &\Leftrightarrow g(\langle x,0\rangle) = g(\langle x,1\rangle). \end{aligned}$$

So we have shown that **RSTC + Quotients** every subset of a set has an indicator, and is accordingly detachable. This proves **(i).**

(ii) By **(i), Detachability** is derivable in **WSTC + Quotients.** This fact easily yields **REM** in **WSTC + Quotients.** Indeed, given restricted $\varphi(x)$, then for any a, the set $b = \{x \in a: \varphi(x)\}$ is a detachable subset of a, from which $\forall x \in a[\varphi(x) \vee \neg\varphi(x)]$ immediately follows. ■

Proof of Theorem 3. It suffices to derive **UEAC** from **AC** in **WST + Quotients.** Assuming $Eq(s, a)$, we use **AC** as in the proof of Theorem 2 to obtain a function $p: a/s \to a$ such that $u = [p(u)]_s$ for all $u \in a/s$. From this we deduce $[x]_s = [p([x]_s)]_s$, whence

(1) $$x \; s \; p([x]_s)$$

for all $x \in a$.

Assuming the antecedent of **UEAC**, viz.,

$$Eq(s,a) \wedge r \subseteq a \times b \wedge Comp(r,s) \wedge \forall x \in a \exists y \in b(x \; r \; y),$$

define the relation $r' \subseteq a/s \times b$ by

$$u \; r' y \Leftrightarrow p(u) \; r \; y \; .$$

Now use **AC** to obtain a function $g: a/s \to b$ for which $\forall u \in a/s (u \; r' \; g(u))$, i.e.

(2) $$\forall u \in a/s (p(u) \; r \; g(u)).$$

Define $f: a \to b$ by

$$f(x) = g([x]_s).$$

Then by (2)

$$\forall x \in a(p([x]_s) \; r \; g([x]_s)).$$

From this, (1) and $Comp(r, s)$ it follows that $\forall x \in a(x \; r \; g([x]_s))$, i.e.

(3) $$\forall x \in a(x \; r \; f(x)).$$

Moreover, for all $x, x' \in a$, we have

$$x \; s \; x' \Rightarrow [x]_s = [x']_s \Rightarrow f(x) = g([x]_s) = g([x']_s) = f(x'),$$

whence *Extn(f, s)*. This, together with (3), establishes the consequent of **UEAC.** ■

We also observe that **Quotients** can be derived within **WST** augmented by the full extensional power set axiom

Extpow $\exists u[Ext(u) \wedge \forall x[x \in u \Leftrightarrow x \subseteq a]]$

So adding *extensional* power sets to **WSTC** yields **REM**[1]**.**

Now recall **AC5:** *unique representatives can be picked from the equivalence classes of any given equivalence relation.* We formulate this as

AC5 $$Eq(s,a) \Rightarrow \exists f[f : a \to a \wedge \forall x \in a(xsf(x)) \wedge \forall x \in a \forall y \in a[xsy \Rightarrow f(x) = f(y)]].$$

Obviously, in **WST**, **Rep** implies **Quotients**. Moreover, the proof of Theorem 2 is easily adapted to show that, in **WST**, **AC5** yields **REM**. In **WST**, **AC + Quotients** entails **Rep**, and, in **WST + Pow**, conversely.

Finally, what about Zermelo's original formulations of the Axiom of Choice **AC1** and **CAC**? The first of these takes the form

AC1 $\forall x \in a[x \subseteq b \wedge \exists y(y \in x)] \Rightarrow \exists f : a \to b\ \forall x \in a(f(x) \in x)].$

This is readily derivable from **AC** in **WST.** If one adds to **WST** the *nonextensional* Power Set Axiom, viz.

Pow $\exists u \forall x[x \in u \Leftrightarrow x \subseteq a],$

then **AC** becomes derivable from **AC1**. Note that while **Extpow** entails **REM, Pow** is logically "harmless", that is, it has no nonconstructive logical consequences such as **LEM.**

The *extensional* version of **AC1**, viz.

EAC1 $$\forall x \in a[x \subseteq b \wedge \exists y(y \in x)] \Rightarrow \exists f : a \to b[Ex(f) \wedge \forall x \in a(f(x) \in x)].$$

[1] cf. Maietti and Valentini [1999].

is derivable in **WST** from **EAC.** In **WST + Pow, EAC** and **EAC1** are equivalent.

Only an extremely weak form of **EAC1,** the *Extensional Axiom of Choice for 2-Doubletons,* is needed to derive **REMS** in **WST**, namely

EACD $$\forall a \subseteq 2 \forall b \subseteq 2[\exists x(x \in a) \wedge \exists x(x \in b) \Rightarrow (\exists f:\{a,b\} \rightarrow 2)[\mathrm{Ex}(f) \wedge f(a) \in a \wedge f(b) \in b]]$$

The argument is similar to that given for Theorem 1(b). Thus given a restricted sentence σ, define

$$a = \{x \in 2 : x = 0 \vee \sigma\}, \quad b = \{x \in 2 : x = 1 \vee \sigma\}.$$

Then $0 \in a \wedge 1 \in b$, so **EACD** gives an extensional function *f*: $\{a,b\} \rightarrow 2$ such that $f(a) \in a \wedge f(b) \in b$, from which we infer $[f(a) = 0 \vee \sigma] \wedge [f(b) = 1 \vee \sigma]$. Applying the distributive law, we then get $\sigma \vee [f(a) = 0 \wedge f(b) = 1]$, whence

(*) $$\sigma \vee f(a) \neq f(b).$$

Now clearly $\sigma \Rightarrow a \approx b$, and from this and the extensionality of *f* *Ext*({*a,b*}) we deduce $\sigma \Rightarrow f(a) = f(b)$, whence $\sigma \Rightarrow f(a) = f(b)$. It follows that $f(a) \neq f(b) \Rightarrow \neg\sigma$, and we conclude from (*) that $\sigma \vee \neg\sigma$, as required.

The second, "combinatorial" form of the Axiom of Choice introduced by Zermelo here takes two forms, the second of which is the extensional version.

CAC $$[\forall x \in a \exists y(y \in x) \wedge \forall x \in a \forall y \in a[\exists z(z \in x \wedge z \in y) \Rightarrow x = y]] \Rightarrow \exists u \forall x \in a \exists! y(y \in x \wedge y \in u)$$

ECAC $$[\forall x \in a \exists y(y \in x) \wedge \forall x \in a \forall y \in a[\exists z(z \in x \wedge z \in y) \Rightarrow x \approx y]] \Rightarrow \exists u \forall x \in a \exists! y(y \in x \wedge y \in u)$$

Clearly **ECAC** implies **CAC**; the former is readily derivable from **EAC** and the latter from **AC.** Since **REMS** is not a consequence of

AC, it cannot, a fortiori, be a consequence of **CAC.** But, like **EAC, ECAC** can be shown to yield **REMS.** We sketch the argument, which is similar to the proof of Thm. 1(b).

Given a restricted sentence σ, define

$$b = \{x \in 2 : x = 0 \vee \sigma\}, \quad c = \{x \in 2 : x = 1 \vee \sigma\}$$

and $a = \{b, c\}$. A straightforward argument shows that a satisfies the antecedent of **ECAC**. So, if this last is assumed, its consequent yields a u with exactly one element in common with b and with c. Writing d and e for these elements, one easily shows that

(*) $$\sigma \vee d \neq e.$$

Now since it is also easily shown that $\sigma \Rightarrow d = e$, it follows that $d \neq e \Rightarrow \neg\sigma$, and this, together with (*) yields $\sigma \vee \neg\sigma$.

*

Within full intuitionistic set theory **AC** implies **LEM** and so retains the complete range of its classical consequences. As we shall see in Chapter VI, however, **ZL** is logically "neutral" in having no nonconstructive consequences within intuitionistic set theory and is also mathematically very weak there. Moving to system **WST**, that is, eliminating the Axiom of Extensionality, amounts, as it were, to levelling the playing field and rendering **AC** and **ZL** equally "impotent".

SOME WEAK FORMS OF AC AND THEIR LOGICAL CONSEQUENCES

Let us term a *weak form* of **AC** any of its inequivalent consequences in classical **ZF**, and a *very weak form* of **AC** a weak form which is provable in **ZF.** We have seen that, in intuitionistic set theory, some very weak forms of **AC** —for instance the assertion that each 2-doubleton has a choice function—imply

LEM. We shall show[1] that there are a number of weak, but not at the same time very weak, forms of **AC** which also imply **LEM** as well as other nonconstructive logical rules. These will come from the theory of posets, and from the theory of distributive lattices and Boolean algebras.

Before we begin our investigations, we need to introduce some more ideas from intuitionistic set theory. In intuitionistic set theory the power set **P**X of any set X is a Heyting algebra under the usual set-theoretic operations: $\cup$ (union), $\cap$ (intersection) and $\mathscr{C}$ (complement). In particular, writing 1 for the one-element set $\{0\}$, **P**1 is a Heyting algebra (see below for a definition) which we shall denote by Ω. Each proposition α of intuitionistic set theory is naturally correlated with the element $\alpha^{\sim} = \{x \in 1: \alpha\}$ of Ω, and each element ω of Ω with the proposition $1 \in \omega$. The correspondence $\alpha \mapsto \alpha^{\sim}$ has the property that $\alpha^{\sim} = \beta^{\sim}$ iff α and β are equivalent. We shall follow the usual practice and identify $\alpha^{\sim}$ with α; in that case the top element 1 of Ω is identified with the identically true proposition **true** and the bottom element $\varnothing$ of Ω with the identically false proposition **false.** These identifications explain why it is customary to call Ω the *algebra of propositions.*

In Chapter III we derived from **ZL** (nonconstructively) the *order extension principle* to the effect that every partial ordering on a set can be extended to a total ordering. We will show that, in intuitionistic set theory, this principle implies the law **Lin** introduced in Chapter **V,** namely $\alpha \Rightarrow \beta \vee \beta \Rightarrow \alpha$ for any propositions α, β.

To prove this, we first observe that if U, $V \subseteq 1$, then

(*) $$(U = 1 \Rightarrow V = 1) \Leftrightarrow U \subseteq V.$$

[1] Bell [1999].

Now suppose that $\leq$ is a partial order on Ω extending $\subseteq$. Then $U \leq 1$ for all $U \subseteq 1$. Now

$$U \leq V \wedge U = 1 \Rightarrow 1 \leq V \Rightarrow V = 1,$$

whence, using (*),

$$U \leq V \Rightarrow (U = 1 \Rightarrow V = 1) \Rightarrow U \subseteq V.$$

We conclude that $\leq$ and $\subseteq$ coincide. Accordingly, if $\subseteq$ could be extended to a total order on Ω, $\subseteq$ would have to be a total order on Ω itself. But this is clearly tantamount to the tassertability of **Lin.**

Next, we require some concepts from the theory of distributive lattices and Boolean algebras. By a *distributive lattice* we shall understand such a lattice $(L, \vee_L, \wedge_L, \leq_L, 0_B, 1_B)$ (again, we shall usually omit the subscript "L") with top and bottom elements 0_L, 1_L. Homomorphisms between distributive lattices in this sense will always be presumed to preserve 0 and 1. A distributive lattice L is a *Heyting algebra* if for each pair a, b of elements of L there is an element of L, which we denote by $a \blacktriangleright b$, such that, for all $x \in L$, $x \wedge a \leq b$ iff $x \leq a \blacktriangleright b$. We write $a \blacktriangleleft\!\blacktriangleright b$ for $(a \blacktriangleright b) \wedge (b \blacktriangleright a)$ and a^* for $a \blacktriangleright 0$. Clearly $a \blacktriangleleft\!\blacktriangleright b = 1$ iff $a = b$. $\blacktriangleright$ and $\blacktriangleleft\!\blacktriangleright$ are called the *implication* and *equivalence* operations, respectively, on H.

We also employ the standard notation and terminology for Boolean algebras. If $(B, \vee_B, \wedge_B, {}^{*_B}, \leq_B, 0_B, 1_B)$ is a Boolean algebra (we shall usually omit the subscript "B"), we write $a \blacktriangleright b$ for $a^* \vee b$ and $a \blacktriangleleft\!\blacktriangleright b$ for $(a \blacktriangleright b) \wedge (b \blacktriangleright a)$. Notice that then, for all $x \in B$, $x \wedge a \leq b$ iff $x \leq a \blacktriangleright b$, so that B is also a Heyting algebra. We write **2** for the initial (two element) Boolean algebra $\{0,1\}$ and **1** for the trivial (one element) Boolean algebra: this is, up to isomorphism, the unique Boolean algebra B in which $0_B = 1_B$. We denote by $\mathcal{B}ool$ the category of Boolean algebras and Boolean homomorphisms. $\mathcal{B}ool$ is a full subcategory of the category of distributive lattices and homomorphisms.

It is easily shown that a Heyting algebra is a Boolean algebra iff it satisfies either of the equivalent identities $x \vee x^* = 1$, $x^{**} \Rightarrow x = 1$. The following are then equivalent: (i) Ω is a Boolean algebra; (ii) the *Law of Excluded Middle*: for any proposition α, α *or* $\neg\alpha$; (iii) the *Law of Double Negation:* for any proposition α, $\neg\neg\alpha \Rightarrow \alpha$.

A Heyting algebra is a *Stone algebra* if it satisfies the identity $x^* \vee x^{**} = 1$, or either of the equivalent identities $(x \wedge y)^* = x^* \vee y^*$, $(x \vee y)^{**} = x^{**} \vee y^{**}$. The following conditions are then equivalent: (i) Ω is a Stone algebra; (ii) for any proposition α, $\neg\alpha$ *or* $\neg\neg\alpha$; (iii) *De Morgan's law*: for any propositions α, β, $\neg(\alpha \ \& \ \beta) \Rightarrow \neg\alpha$ *or* $\neg\beta$; (iv) for any propositions α, β, $\neg\neg(\alpha$ *or* $\beta) \Rightarrow \neg\neg\alpha$ *or* $\neg\neg\beta$.

If Y is a subset of a set X, write $\mathcal{C}Y$ for the complement $\{x \in X: x \notin Y\}$ of Y.Y is called *stable* if $\mathcal{C}\mathcal{C}Y = Y$, that is, if, for any $x \in X$ $\neg\neg(x \in Y) \Rightarrow x \in Y$; it is *complemented* if $Y \cup \mathcal{C}Y = X$, that is, if, for any $x \in X$, either $x \in Y$ or $\neg\ x \in Y$: clearly any complemented set is stable (but not conversely). For any set X, the families **C**X and **S**X of complemented and stable subsets, respectively, of X form Boolean algebras: the operations on the former are the usual set-theoretical ones; the same is true for the latter with the exception of $\vee$, which is defined to be the *double complement* of the union. We write $\Omega_{\neg\neg}$ for **S**1; and clearly **C**1 is (isomorphic to) the initial Boolean algebra **2**.

A *filter* (resp., *ideal*) in a distributive lattice L is a subset F (resp., I) such that $1 \in F$, $x, y \in F \Rightarrow x \wedge y \in F$, $x \in F \ \& \ x \leq y \Rightarrow y \in F$ (resp. $0 \in I$, $x, y \in I \Rightarrow x \vee y \in I$, $x \in I \ \& \ y \leq x \Rightarrow y \in I$.) A filter F (ideal I) is *proper* if $0 \notin F$ ($1 \notin I$); clearly a distributive lattice is trivial iff it contains no proper filters (or no proper ideals). A filter F (ideal I) in L is *prime* if it is proper and satisfies the condition $x \vee y \in F \Rightarrow x \in F$ or $y \in F$ ($x \wedge y \in I \Rightarrow x \in I$ or $y \in I$): if L is a Boolean algebra, this is equivalent to the condition that, for any x, $x \in F$ or $x^* \in F$ ($x \in I$ or $x^* \in I$). Note that it follows immediately from this that both *prime filters and prime ideals in Boolean algebras are complemented.* It follows in turn that for each Boolean algebra B, there is a natural correspondence between prime filters (or ideals) and homomorphisms $B \to \mathbf{2}$: each prime filter P in B is correlated with the homomorphism $h: B \to \mathbf{2}$ defined by $h(x) = 1$ iff $x \in P$, and each homomorphism $h: B \to \mathbf{2}$ with the prime filter $h^{-1}[1]$. A filter (ideal) is an *ultrafilter* (*maximal ideal*) if it is proper and maximal with respect to that property. It is readily shown that a proper filter F is an ultrafilter iff it satisfies the condition $\forall x[\forall y \in F(x \wedge y \neq 0 \Rightarrow x \in F]$, and that a proper ideal I is maximal iff it satisfies the condition $\forall x[\forall y \in I(x \vee y \neq 1 \Rightarrow x \in I]$, In a Heyting algebra these conditions are easily shown to be equivalent to $\forall x[x \notin F \Rightarrow x^* \in F]$ and $\forall x[x \notin I \Rightarrow x^* \in I]$. We note that *ultrafilters* (and maximal ideals) *in distributive lattices are stable*. For it is readily shown that the double complement of a proper filter is a proper filter; thus, if U is an ultrafilter, $\mathscr{C}\mathscr{C}U$ is a proper filter containing, and so identical with, U.

Recall that the classical *Stone Representation Theorem* for Boolean algebras asserts that every Boolean algebra is isomorphic to a subalgebra of $\mathbf{P}S$ for some set S. In a constructive context, we observe that since every member of a Boolean algebra of subsets of a set is obviously complemented, in the statement of this theorem "$\mathbf{P}S$" may be replaced by "$\mathbf{C}S$".

We call a distributive lattice (in particular, a Boolean algebra) *semisimple* if the intersection of the family of all its prime filters is {1}. A Boolean algebra C is said to be a *cogenerator* in $\mathcal{B}ool$ if it has the following property: for any pair of parallel morphisms $f, g: A \to B$ in $\mathcal{B}ool$, if $h \circ f = h \circ g$ for all $h: B \to C$, then $f = g$.

We shall need the following result of Peremans [1957]:

(**Per**). *It is constructively provable that any distributive lattice can be embedded in a Boolean algebra.*

Theorem 1. *The following assertions are constructively equivalent.*

(i) *The Stone Representation Theorem for Boolean algebras;*

(ii) *the Stone Representation Theorem for distributive lattices: any distributive lattice is isomorphic to a lattice of subsets of a set;*

(iii) *any distributive lattice is semisimple;*

(iv) *any Boolean algebra is semisimple;*

(v) *the initial Boolean algebra* **2** *is a cogenerator in* $\mathcal{B}ool$.

Proof. **(i) ⇔ (ii).** One direction is obvious. By **Per**, any distributive lattice is constructively embeddable in a Boolean algebra, so **(i) ⇒ (ii)** follows immediately.

(ii) ⇒ (iii). Assume **(ii)**; then any distributive lattice L may be considered a sublattice of $\mathbf{P}S$ for some set S. For any $x \in S$, $F_x = \{X \in L: x \in X\}$ is a prime filter; if $X \in \bigcap\{F_x: x \in S\}$, then $x \in X$ for all $x \in S$, whence $X = S$. Therefore $\bigcap\{F_x: x \in S\} = \{S\}$, and L is semisimple.

Conversely, assume (iii). Given a distributive lattice L, let S be the set of all prime filters in L, and define $h: L \to \mathbf{P}S$ by $h(x) = \{F \in S: x \in F\}$. It is easy to see that h is a homomorphism; the semisimplicity of L implies that h is injective. Hence **(ii).**

(i) ⇒ (iv). The proof of this is similar to that of **(ii) ⇒ (iii).**

(iv) ⇒ (v). Assume **(iv)** and suppose that $f, g: A \to B$ are such that if $h \circ f = h \circ g$ for all $h: B \to \mathbf{2}$. Then for all $h: B \to \mathbf{2}$ and $x \in A$ we have $h(f(x)) = h(g(x))$ so that $1 = h(f(x)) \leftrightarrow h(g(x)) = h(f(x)) \leftrightarrow h(g(x))$. Under the natural correspondence between homomorphisms $B \to \mathbf{2}$ and prime filters in B, this means that $f(x) \leftrightarrow g(x)$ is contained in every prime filter in B. Since B is semisimple, it follows that $f(x) \leftrightarrow g(x) = 1$, so that $f(x) = g(x)$ for every $x \in A$, i.e. $f = g$. Hence **(v).**

Conversely, assume **(v).** Consider the 4-element Boolean algebra

4 =

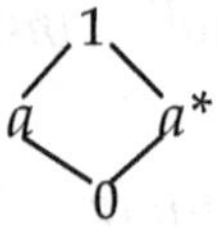

For any Boolean algebra B, each homomorphism $\mathbf{4} \to B$ is uniquely determined by the image of a, which can be an arbitrary element b of B. Denote this homomorphism by $b^{\sim}$. Suppose now that every prime filter in B contains b. Then, under the natural correspondence between prime filters in B and homomorphisms $B \to \mathbf{2}$, this means that $h(b) = h(1)$, whence $h \circ b^{\sim} = h \circ 1^{\sim}$ for all $h: B \to \mathbf{2}$. By (v), $b^{\sim} = 1^{\sim}$, so that $b = 1$, and B is semisimple. ■

Theorem 2. *Any of* **(i) - (v)** *of Thm. 1 constructively implies that* Ω *is a Boolean algebra.*

Proof. Let us assume, for instance, **(iv).** For each Boolean algebra B, let Prim(B) be the set of prime filters in B. Then $\bigcap$Prim(B) = {1} and we have

$$(*) \qquad \text{Prim}(B) = \varnothing \Rightarrow B \text{ is trivial.}$$

For if B is trivial, it has no proper filters, so that Prim(B) = ∅. Conversely, if Prim (B) = ∅, then $\{1\} = \bigcap\text{Prim}(B) = \bigcap\varnothing = B$, so that B is trivial.

Now let α be any proposition, and define

$$B_\alpha = \{\omega \in \Omega : \omega = \alpha \textit{ or } \omega = \textbf{true}\}.$$

This is easily shown to be a Boolean algebra in which $0 = \alpha$, $1 =$ **true**, meets are conjunctions, joins are disjunctions, and the complement of ω is $(\omega \Rightarrow \alpha)$. Clearly

(**) $$B_\alpha \text{ is trivial} \Leftrightarrow \alpha.$$

Putting (*) and (**) together, we see that

$$\alpha \Leftrightarrow \mathrm{Prim}(B_\alpha) = \varnothing \Leftrightarrow \neg\exists X.\ X \in \mathrm{Prim}(B_\alpha).$$

Thus α is equivalent to a negated statement, so that $\neg\neg\alpha \Rightarrow \alpha$. Since α was arbitrary, it follows that Ω is a Boolean algebra. ■

Thm. 1 can also be stated and proved, in a similar way, for *nontrivial* Boolean algebras and distributive lattices. However, the proof that any one of the correspondingly weakened versions of conditions **(i)** - **(v)** implies that Ω is a Boolean algebra differs from the proof of Thm. 2, as witness:

Theorem 3. *The assertion* any nontrivial Boolean algebra is semisimple *constructively implies that Ω is a Boolean algebra.*

Proof. Let B be a semisimple Boolean algebra. Then $\{1\}$, as the intersection of prime filters, is the intersection of complemented sets and is therefore (as is easily seen), stable. So the premise of the present Theorem implies that $\{1\}$ is a stable subset of every nontrivial Boolean algebra. Now, by **Per** , Ω is constructively embeddable in a – necessarily nontrivial – Boolean algebra B, so we may consider Ω as a subset of B. Then $\{1\} = \{\textbf{true}\}$ is a stable subset of B and hence also of Ω. But the stability of $\{\textbf{true}\}$ in Ω is obviously equivalent to the assertion that it be a Boolean algebra. ■

Classically, the Stone Representation Theorem is equivalent to the assertion that **2** be *injective*[1] in $\mathcal{B}ool$. This equivalence is not

[1] Recall that a Boolean algebra C is *injective* (in $\mathcal{B}ool$) if any homomorphism to C from a subalgebra of any Boolean algebra B can be extended to the whole of B.

constructively valid, since while the former can hold only when Ω is a Boolean algebra, the latter can be true even when Ω is merely a Stone algebra. To see that the injectivity of **2** implies that Ω is a Stone algebra, observe that from this assumption it follows that the Boolean algebra $\Omega_{\neg\neg}$ must have a homomorphism to **2**, and hence must also contain a prime filter. Since {**true**} is the only proper filter in $\Omega_{\neg\neg}$, it must be both prime and an ultrafilter. Then $\mathscr{C}\mathscr{C}$\{**true**\} = {**true**} is prime, that is, for α, β in $\Omega_{\neg\neg}$,

$$\alpha \vee_{\neg\neg} \beta \Rightarrow \alpha \textit{ or } \beta,$$

where $\vee_{\neg\neg}$ is the join calculated in $\Omega_{\neg\neg}$. Since (as is easily verified) $\alpha \vee_{\neg\neg} \beta = \neg\neg(\alpha \textit{ or } \beta)$, we infer

$$\neg\neg(\alpha \textit{ or } \beta) \Rightarrow \alpha \textit{ or } \beta,$$

Now for arbitrary α, β in Ω, $\neg\neg\alpha$, $\neg\neg\beta$ are in $\Omega_{\neg\neg}$, so it follows that

$$\neg\neg(\alpha \textit{ or } \beta) \Rightarrow \neg\neg(\neg\neg\alpha \textit{ or } \neg\neg\beta) \Rightarrow \neg\neg\alpha \textit{ or } \neg\neg\beta,$$

and therefore Ω is a Stone algebra.

In conclusion, we show that the injectivity of **2** is constructively equivalent to a number of familiar results in the theory of Boolean algebras.

Theorem 4. *The following are constructively equivalent (and each implies that Ω is a Stone algebra).*

(i) *For any Boolean algebra B and any $x \neq 0$ in B there is $h: B \rightarrow$ **2** such that $h(x) = 1$.*

(ii) *For any Boolean algebra B and any $x \neq 0$ in B there is a prime filter in B containing x.*

(iii) *Any nontrivial Boolean algebra contains a prime filter.*

(iv) *Each proper filter in a Boolean algebra is contained in a prime filter.`*

(v) **2** *is injective in $\mathscr{B}ool$.*

(vi) *For any Boolean algebra B, there is a set S and a homomorphism h: B → **P**S such that, for any x ∈ B, x ≠ 0 → h(x) is inhabited.*[1]

Proof. (i) ⇒ (ii) ⇒ (iii) are all obvious.

(iii) ⇒ (iv). Assume **(iii)** and let F be a proper filter in a Boolean algebra B. Then the quotient B/F is nontrivial and so contains a prime filter P. The inverse image $\pi^{-1}[P]$ of P under the canonical homomorphism $\pi: B \to B/F$ is easily seen to be a prime filter in B containing F.

(iv) ⇒ (v). Assume **(iv),** let A a subalgebra of a Boolean algebra B, and let h be a homomorphism of A to **2**. Then $h^{-1}[1]$ is a (prime) filter in A in turn generating a proper filter in B which, by **(iv),** is contained in a prime filter P in B. The homomorphism $B \to \mathbf{2}$ naturally corresponding to P is an extension of h.

(v) ⇒ (iii). Assume (v) and let B be a nontrivial Boolean algebra. Then **2** may be considered a subalgebra of b and the identity homomorphism $\mathbf{2} \to \mathbf{2}$ has an extension to B, giving rise to a naturally correlated prime filter in B.

(iv) ⇒ (vi). Assume **(iv),** and let S be the set of prime filters in a given Boolean algebra B. Define $h: B \to \mathbf{P}S$ by $h(x) = \{F \in S: x \in F\}$. This h is a homomorphism; if $x \neq 0$ in B, then x generates a proper filter which is contained in a prime filter P. Then $P \in h(x)$ and $h(x) \not\equiv \varnothing$. Hence **(vi).**

(vi) ⇒ (ii). Assume **(vi)** and the data of **(ii).** If $a \neq 0$ in B, then $h(a)$ is inhabited, so there is an element $s \in h(a)$. Then $\{x \in B: s \in h(x)\}$ is a prime filter in B containing a. (ii) follows. ■

[1] We recall that, in constructive mathematics, a set X is said to be *inhabited* if $\exists x.\, x \in X$.

VI
The Axiom of Choice in Category Theory, Topos Theory and Local Set Theory

AC IN CATEGORICAL CLOTHING

Some significant recent work on the foundational role **of AC** has arisen in connection with *category theory.* **AC** admits a natural category-theoretic formulation in its version **AC4.** Thus a category $\mathscr{C}$ is said to satisfy **AC** if each epic arrow has a right inverse or a *section*, that is, given any epic arrow $f: A \twoheadrightarrow B$, there is an arrow $g: B \rightarrow A$ for which $fg = 1_B$. It is readily seen that a category satisfies **AC** precisely when each of its objects is projective.

Another category-theoretic formulation **of AC** is associated with version **AC4*.** Although in classical set theory **AC4** and**AC4*** are equivalent, within a category the latter is, in general, stronger than the former. Accordingly, given a category $\mathscr{C}$ with a terminal object 0, we shall say that $\mathscr{C}$ satisfies the *Strong Axiom of Choice* (which we shall abbreviate to **SAC**) if it satisfies the categorical version of **AC4***, namely,

for any object $X \not\cong 0$ *and any arrow* $f: X \rightarrow Y$, *there is an arrow* $g: Y \rightarrow X$ *such that* $fgf = f$.

Now it is most unusual for a category to satisfy **AC**, that is, for all of its objects to be projective. In most categories projective objects are quite special. Here is a table in which the projective objects within some familiar categories are identified[1]:

[1] It should be noted that the use of **AC** is required in all but the first two lines of this table.

Category	Projective objects
TOPOLOGICAL SPACES	DISCRETE SPACES
POSETS	TRIVIALLY ORDERED SETS[1]
ABELIAN GROUPS	TORSION-FREE GROUPS
GROUPS	RETRACTS OF FREE GROUPS
BOOLEAN ALGEBRAS	RETRACTS OF FREE BOOLEAN ALGEBRAS
COMPACT HAUSDORFF SPACES	EXTREMALLY DISCONNECTED SPACES[2]

In fact, the only "natural" category which could *possibly* satisfy **AC** is the category ***Set*** of sets[3]. The reason for this is not difficult to find. For a set, in the mathematical sense, is presumed to consist of a plurality of unrelated elements which have been purged of all intrinsic qualities *aside from the quality which distinguishes each element from the rest.* A set in this sense—let us call it a *pure* set[4]— is accordingly an image of pure discreteness, an embodiment of raw difference; in short, it is an assemblage of unchanging, featureless, *but nevertheless distinct* "dots" or "motes"[5]. The sole intrinsic attribute of a set conceived in this way is the number of its elements. Given this, it follows that there are no constraints on the correspondences, the *mappings* between pure sets: these mappings can be completely arbitrary, they are not required to be continuous, or order-preserving, or indeed to preserve any structure at all. It is this feature of pure sets which

[1] A partially ordered set is *trivially* ordered if its ordering coincides with the identity relation.

[2] A topological space is *extremally disconnected* if the closure of any of open subset is open.

[3] Aside from minor variations such as categories of Boolean-valued sets: see below.

[4] For further discussion of pure sets, see Bell [2006a], where they are called "abstract" sets.

[5] Perhaps also as "marks" or "strokes" in Hilbert's sense.

makes **AC** a natural principle in *Set*. Thus, in the figure below, the choice of a section *s* of the epic map *p* can be made on purely combinatorial grounds, since *no constraint whatsoever has*

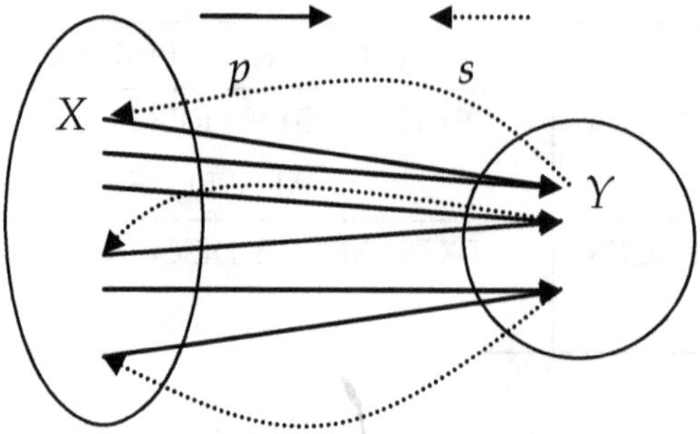

been placed on s (aside, of course, from the fact that it must be a right inverse of *p*). So we see that **AC4** holds in *Set*; a similar figure shows that **AC4*** (or **SAC**) also holds there.

Now as soon as one moves from *Set* to a category whose objects carry some nontrivial structure which has to be preserved by its maps, one cannot simply produce a section to an epic map in the "combinatorial" manner just used for pure sets: any such section must also preserve the structure carried by the objects of the category, and this may simply not be possible. Consider, for instance, the map $p: P \to Q$ in the category *Poset* of partially ordered sets and order-preserving maps as illustrated below:

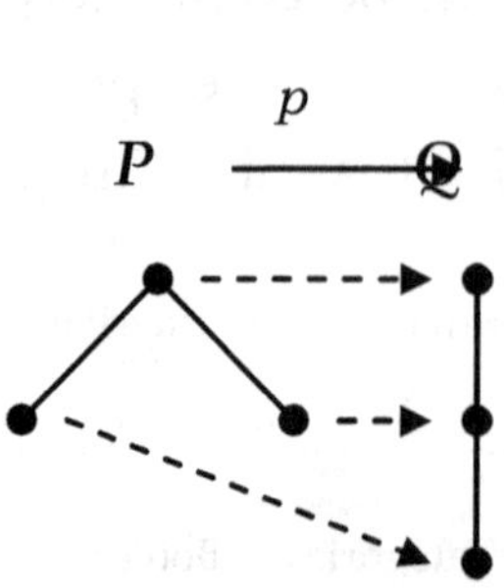

It should be clear that p cannot have a section which is order-preserving.

As another example, consider the continuous map $p: S^1 \to S^1$ given by $p(e^{i\theta}) = e^{2i\theta}$. (Here S^1 is the unit circle, regarded as a subset of the complex plane.) The map p —the "double covering" map of S^1 (depicted below) —is an epic arrow in the category *Top* of topological spaces. But it has no section in

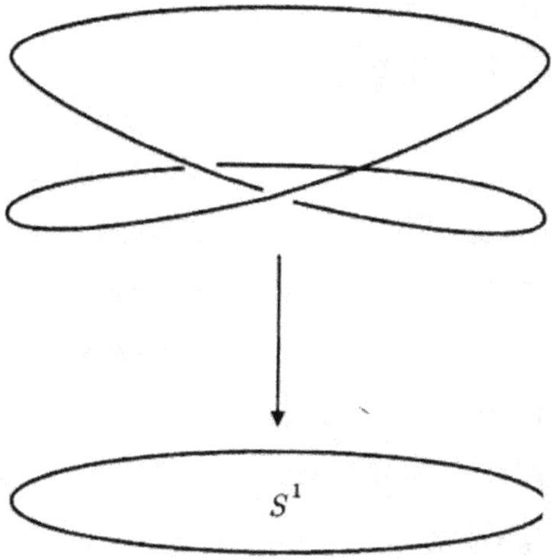

Top, for any such section would have to be a homemorphism of S^1 onto a half-circle, which is impossible since S^1 , but no half-circle, remains connected when a single point is removed . We see, then, that **AC** will typically fail when structure is imposed on pure sets. Another move that will cause **AC** to fail is to subject pure sets to *variation*. The objects of *Set* have been conceived as pluralities which, in addition to being discrete, are also *static* or *constant* in the sense that their elements undergo no change.

There are a number of natural category-theoretic approaches to bringing variation into the picture. For example, we can introduce a simple form of *discrete* variation by considering as objects *bivariant sets*, that is, maps $F: X_0 \longrightarrow X_1$ between pure sets. Here we think of X_0 as the "state" of the bivariant set F at stage 0, or "then", and X_1 as its "state" at stage 1, or "now". The bivariant set may be thought of having undergone, via the "transition" F, a change from what it was then (X_0) to what it is now (X_1). Any element x of X_0, that is, of F "then" becomes the element Fx of X_0 "now". Pursuing this metaphor, two elements "then" may become one "now" (if F is not monic), or a new element may arise "now", but because F is a map, no element "then" can split into two or more "now" or vanish altogether.

The appropriate maps between bivariant sets are pairs of maps between their respective states which are compatible with transitions. Thus a map from $F: X_0 \longrightarrow X_1$ to $G: Y_0 \longrightarrow Y_1$ is a pair of maps $h_0: X_0 \longrightarrow Y_0$, $h_2: X_1 \longrightarrow Y_1$ for which $G \circ h_1 = h_2 \circ F$. Bivariant sets and maps between them defined in this way form the category $\mathcal{Biv}$ of bivariant sets.

Now AC fails in $\mathcal{Biv}$. Indeed, it is easily checked that the epic arrow from the identity map on {0, 1} to the map {0, 1} → {0} depicted below has no section in $\mathcal{Biv}$:

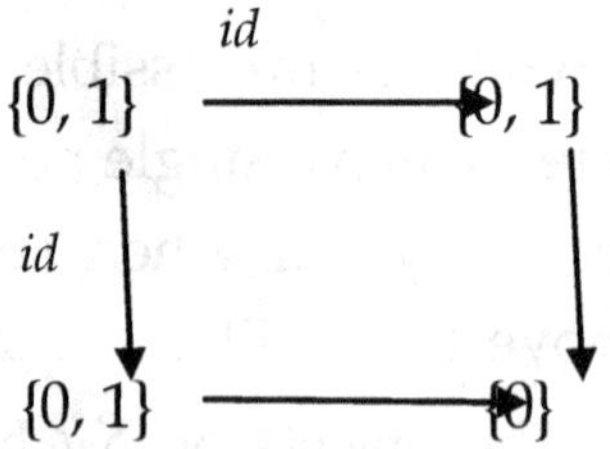

Thus **AC** is incompatible with even the most rudimentary sort of discrete variation of pure sets. Now pure sets can also be subjected to *continuous* variation. This can be achieved in the first instance by considering, in place of pure sets, *bundles over*

topological spaces. Here a *bundle* over a topological space X is a continuous map p from some topological space Y to X. If we think of the space Y as the union of all the "fibres" $A_x = p^{-1}(x)$ for $x \in X$, and A_x as the "value" at x of the pure set A, then the bundle p itself may be conceived as the *pure set A varying continuously over* X. A *map* $f: p \to p'$ between two bundles $p: Y \to X$ and $p': Y' \to X$ over X is a continuous map $f: Y \to Y'$ respecting the variation over X, that is, satisfying $p' \circ f = p$. Bundles over X and maps between them form a category $\mathcal{Bun}(X)$, the *category of bundles* over X.

While categories of bundles represent the idea of continuously varying sets in a weak sense, as categories they do not resemble ***Set*** sufficiently[1] to be taken as suitable generalizations of ***Set*** embodying such variation. To obtain these, we confine attention to special sorts of bundles known as *displayed spaces*[2]. A bundle $p : Y \to X$ over X is called a *displayed space* over X when p is a *local homeomorphism* in the following sense: to each $a \in Y$ there is an open neighbourhood U of a such that pU is open in X and the restriction of p to U is a homeomorphism $U \to pU$. The domain space of a displayed space over X then "locally resembles" X in the same sense as a differentiable manifold locally resembles Euclidean space. *Categories of displayed spaces provide the appropriate generalizations of the category of pure sets to allow for continuous variation,* and the term *continuously varying set* is taken to be synonymous with the term *displayed space.* We write $\mathcal{Esp}(X)$

---

[1] To be precise, in general they fail to satisfy the topos axioms. These latter are given in Appendix II.

[2] French "espace étalé".

for the category of displayed spaces over X with bundle maps between them.

If we take X to be a space consisting of a single point, a displayed space over X is just a discrete space, so that the category of sheaves over X is essentially the category of pure sets. In other words, a pure set varying continuously over a one-point space is just a (constant) pure set. In this way arresting continuous variation leads back to constant discreteness[1].

In general, **AC** fails in categories of displayed spaces, showing that it is also incompatible with continuous variation. This is most easily seen by considering the "double covering" map p: $S^1 \rightarrow S^1$ described above. The map p is easily seen to be a local homeomorphism, and the fact that it has no continuous section implies that the natural epic map in $\mathcal{E}sp(S^1)$ from p: $S^1 \rightarrow S^1$ to the identity map $S^1 \rightarrow S^1$ (the terminal object of $\mathcal{E}sp(S^1)$) has no section, so that **AC** fails in $\mathcal{E}sp(S^1)$.

Having demonstrated that **AC** is incompatible with structure, and with variation, both discrete and continuous, we conclude that it can hold only within a realm of static and structureless objects—that is, the realm of pure sets. Indeed, as we shall demonstrate in the final section of this chapter, in a certain sense **AC** *characterizes* the realm of pure sets.

---

[1] Observe that had we chosen categories of bundles to represent continuous variation, the corresponding arresting of variation would lead, not to the category of abstract sets—constant discreteness—but to the category of topological spaces—constant continuity. This is another reason for not choosing bundle categories as the correct generalization of the category of pure sets to incorporate continuous variation.

LOCAL SET THEORIES

There are certain evident *basic axioms* satisfied by the category $\mathcal{S}et$ of pure sets:

1. There is a 'terminal' object 1 such that, for any object X, there is a unique arrow $X \to 1$
2. Any pair of objects A, B has a Cartesian product $A \times B$.
3. For any pair of objects A, B one can form the 'exponential' object B^A of all maps $A \to B$.
4. There is a "truth value" object Ω such that for each object X a natural correspondence exists between subobjects (subsets) of X and arrows $X \to \Omega$. (In $\mathcal{S}et$, one may take Ω to be the set $2 = \{\varnothing, 1\}$.)
5. 1 is not isomorphic to $\varnothing$.
6. The Axiom of Infinity: there exists an object X for which X is isomorphic to $X + 1$. A pure set X is said to be *infinite* if there exists an isomorphism between X and the set $X + 1$ obtained by adding one additional "dot" to X.
7. "Extensionality" principle: for any objects A, B and any pair of arrows $A \xrightarrow{f} B, A \xrightarrow{g} B$, if $fh = gh$ for every arrow $1 \xrightarrow{h} A$, then $f = g$. This says that each object satisfies the axiom of extensionality in the sense that its identity as a domain is entirely determined by its "elements".

A category satisfying axioms **1. – 5.** (suitably formulated in categorical language) is called a (nondegenerate) *topos*[1]. Accordingly $\mathcal{S}et$ is an extensional topos satisfying both the Axiom of Infinity and **SAC**. $\mathcal{B}iv$ and $\mathcal{E}sp(X)$ are toposes, but as we have seen they do not satisfy **AC**. There are numerous others. As we

[1] For the technical definition of a topos, see Appendix II below.

shall see, there is a sense in which the fact that $\mathcal{S}et$ satisfies **SAC** *characterizes* it as a topos.

The role played by **AC** in topos theory is brought out most clearly by presenting the latter in terms of the intuitionistic type theories with which toposes are associated. These are known as *local set theories.*[1] We shall sketch enough of the development of local set theories to enable **AC** to become visible in that setting.

A local set theory is a type-theoretic system built on the same primitive symbols =, ∈, {:} as classical set theory, in which the set-theoretic operations of forming products and powers of types can be performed, and which in addition contains a "truth value" type acting as the range of values of "propositional functions" on types. A local set theory is determined by specifying a collection of *axioms* formulated within a *local language* defined as follows.

A *local language* $\mathscr{L}$ has the following *basic symbols:*

- **1** (*unit type*) **Ω** (*truth value type* or *type of propositions*)
- **S, T, U,...** (*ground types*: possibly none of these)
- **f, g, h,...** (*function symbols*: possibly none of these)
- $x_{\mathbf{A}}, y_{\mathbf{A}}, z_{\mathbf{A}}, \ldots$ (*variables of each type* **A**, where a *type* is as defined below)
- ★ (*unique entity of type* **1**)

The *types* of $\mathscr{L}$ are defined recursively as follows:

- **1, Ω** are types
- any ground type is a type
- $\mathbf{A}_1 \times \ldots \times \mathbf{A}_n$ is a type whenever $\mathbf{A}_1, \ldots, \mathbf{A}_n$ are, where, if $n = 1$, $\mathbf{A}_1 \times .. \times \mathbf{A}_n$ is $\mathbf{A}_1$, while if $n = 0$, $\mathbf{A}_1 \times .. \times \mathbf{A}_n$ is **1** (*product types*)

[1] For a fuller account of toposes and local set theories, see Bell [1988].

- **PA** is a type whenever **A** is (*power types*)

Each function symbol **f** is assigned a *signature* of the form **A → B**, where **A, B** are types; this is indicated by writing **f: A → B**.

Terms of $\mathscr{L}$ and their associated *types* are defined recursively as follows. We write τ: **A** to indicate that the term τ has type **A.**

Term: type	Proviso
★: **1**	
$x_{\mathbf{A}}$: **A**	
f(τ): **B**	**f: A → B** τ: **A**
<τ₁, ..., τₙ>: $\mathbf{A}_1 \times \ldots \times \mathbf{A}_n$, where $<\tau_1, \ldots, \tau_n>$ is τ_1 if $n = 1$, and ★ if $n = 0$.	τ_1: $\mathbf{A}_1, \ldots, \tau_n$: $\mathbf{A}_n$
$(\tau)_i$: $\mathbf{A}_i$ where $(\tau)_i$ is τ if $n = 1$	τ: $\mathbf{A}_1 \times \ldots \times \mathbf{A}_n$, $1 \leq i \leq n$
$\{x_{\mathbf{A}}: \alpha\}$: **PA**	α : **Ω**
σ = τ: **Ω**	σ, τ of same type
σ ∈ τ: Ω	σ: **A,** τ: **PA** for some type **A**

Terms of type **Ω** are called *formulas, propositions,* or *truth values.* Notational conventions we shall adopt include:

$\omega, \omega', \omega''$	variables of type Ω
α, β, γ	formulas
x, y,,z ...	$x_{\mathbf{A}}, y_{\mathbf{A}}, z_{\mathbf{A}}$...
τ(*x*/σ) or τ(σ)	result of substituting σ at each free occurrence of *x* in τ: an occurrence of *x* is *free* if it does not appear within {*x*: α}
α ⇔ β	α = β
Γ : α	sequent notation; Γ a finite set of formulas
: α	∅: α

A term is *closed* if it contains no free variables; a closed term of type **Ω** is called a *sentence*.

The *basic axioms* in $\mathscr{L}$ are as follows:

Unity $: x_1 = \star$

Equality $x = y, \alpha(z/x) : \alpha(z/y)$ (x, y free for z in α)

Products $: (<x_1, ..., x_n>)_i = x_i$

$: x = <(x)_1, ..., (x)_n>$

Comprehension $: x \in \{x: \alpha\} \Leftrightarrow \alpha$

The *rules of inference* in $\mathscr{L}$ are the following:

Thinning

$$\frac{\Gamma : \alpha}{\beta, \Gamma : \alpha}$$

Restricted Cut

$$\frac{\Gamma : \alpha \quad \alpha, \Gamma : \beta}{\Gamma : \beta}$$ (any free variable of α free in Γ or β)

Substitution

$$\frac{\Gamma : \alpha}{\Gamma(x/\tau) : \alpha(x/\tau)}$$ (τ free for x in Γ and α)

Extensionality

$$\frac{\Gamma : x \in \sigma \Leftrightarrow x \in \tau}{\Gamma : \sigma = \tau}$$ (x not free in Γ, σ, τ)

Equivalence

$$\frac{\alpha, \Gamma : \beta \quad \beta, \Gamma : \alpha}{\Gamma : \alpha \Leftrightarrow \beta}$$

These axioms and rules of inference yield a system of *natural deduction* in $\mathscr{L}$. If S *is* any collection of sequents in $\mathscr{L}$, we say that the sequent $\Gamma : \alpha$ is *deducible from* S, and write $\Gamma \vdash_S \alpha$ provided there is a derivation of $\Gamma : \alpha$ using the basic axioms, the sequents in S, and the rules of inference. We shall also write $\Gamma \vdash_S \alpha$ for $\Gamma \vdash_{\varnothing} \alpha$ and $\vdash_S \alpha$ for $\varnothing \vdash_S \alpha$. We say that α is *S-derivable* if $\vdash_S \alpha$.

A *local set theory* in $\mathscr{L}$ is a collection S of sequents closed under deducibility from S. Any collection of sequents S *generates* the local set theory S^* comprising all the sequents deducible from S. The local set theory in $\mathscr{L}$ generated by $\varnothing$ is called *pure* local set theory in $\mathscr{L}$.

The *logical operations* in $\mathscr{L}$ are defined as follows:

Logical Operation	**Definition**
$\top$ (true)	$\bigstar = \bigstar$
$\alpha \wedge \beta$	$<\alpha, \beta> = <\top, \top>$
$\alpha \Rightarrow \beta$	$(\alpha \wedge \beta) \leftrightarrow \alpha$
$\forall x\, \alpha$	$\{x : \alpha\} = \{x : \top\}$
$\perp$ (false)	$\forall \omega.\ \omega$
$\neg\alpha$	$\alpha \rightarrow \perp$
$\alpha \vee \beta$	$\forall \omega[(\alpha \Rightarrow \omega \wedge \beta \Rightarrow \omega) \Rightarrow \omega]$
$\exists x\, \alpha$	$\forall \omega[\forall x(\alpha \Rightarrow \omega) \Rightarrow \omega]$

We write $x \neq y$ for $\neg(x = y)$, and $x \notin y$ for $\neg(x \in y)$. We also define the *unique existential quantifier* $\exists!$ in the familiar way, namely,

$$\exists! x\alpha \;\equiv\; \exists x[\alpha \wedge \forall y(\alpha(x/y) \Rightarrow x = y).$$

It can be shown[1] that the logical operations on formulas just defined satisfy the axioms and rules of intuitionistic logic.

[1] Bell [1988].

A local set theory S is said to be *consistent* if it is not the case that $\vdash_S \bot$.

SET THEORY IN A LOCAL LANGUAGE

Now we introduce the concept of *set* in a local language. A *set-like* term is a term of power type; a *closed* set-like term is called an ($\mathscr{L}$-) *set*. We shall use upper case italic letters $X, Y, Z, \ldots$ for sets, as well as standard abbreviations such as $\forall x \in X.\alpha$ for $\forall x(x \in X \Rightarrow \alpha)$. If X is an ($\mathscr{L}$-) set, then X is of type **PA** for some type **A**; a closed term **a** of type **A** such that $\vdash_S \mathbf{a} \in \mathbf{X}$ is called an *S-element* of X.

Set-theoretic *operations* and *relations* on $\mathscr{L}$- sets are defined as follows. Note that in the definitions of $\subseteq$, $\cap$, and $\cup$, X and Y *must be of the same type*:

Operation	Definition
$\{x \in X : \alpha\}$	$\{x : x \in X \wedge \alpha\}$
$X \subseteq Y$	$\forall x \in X.\ x \in Y$
$X \cap Y$	$\{x : x \in X \wedge x \in Y\}$
$X \cup Y$	$\{x : x \in X \vee x \in Y\}$
$x \notin X$	$\neg(x \in X)$
$U_{\mathbf{A}}$ or A	$\{x_{\mathbf{A}} : \top\}$
$\varnothing_{\mathbf{A}}$ or $\varnothing$	$\{x_{\mathbf{A}} : \bot\}$
$E - X$	$\{x : x \in E \wedge x \notin X\}$
PX	$\{u : u \subseteq X\}$
$\bigcap U$ $(U : \mathbf{PPA})$	$\{x : \forall u \in U.\ x \in u\}$
$\bigcup U$ $(U : \mathbf{PPA})$	$\{x : \exists u \in U.\ x \in u\}$
$\bigcap_{i \in I} X_i$	$\{x : \forall i \in I.\ x \in X_i\}$
$\bigcup_{i \in I} X_i$	$\{x : \exists i \in I.\ x \in X_i\}$
$\{\tau_1, \ldots, \tau_n\}$	$\{x : x = \tau_1 \vee \ldots \vee x = \tau_n\}$
$\{\tau : \alpha\}$	$\{z : \exists x_1 \ldots \exists x_n (z = \tau \wedge \alpha)\}$

$X \times Y$	$\{<x,y>: x \in X \wedge y \in Y\}$
$X + Y$	$\{<\{x\},\varnothing>: x \in X\} \cup \{<\varnothing,\{y\}.: y \in Y\}$
$Fun(X, Y)$ or Y^X	$\{u: u \subseteq X \times Y \wedge \forall x{\in}X\, \exists! y{\in} Y.\ <x,y> \in u\}$

The following facts concerning the set-theoretic operations and relations may now be established as straightforward consequences of their definitions:

(i) $\vdash X = Y \Leftrightarrow \forall x(x \in X \leftrightarrow x \in Y)$

(ii) $\vdash X \subseteq X,\quad \vdash (X \subseteq Y \wedge Y \subseteq X) \Rightarrow X = Y,$

$\vdash (X \subseteq Y \wedge Y \subseteq Z) \Rightarrow X \subseteq Z$

(iii) $\vdash Z \subseteq X \cap Y \Leftrightarrow Z \subseteq X \wedge Z \subseteq Y$

(iv) $\vdash X \cup Y \subseteq Z \Leftrightarrow X \subseteq Z \wedge Y \subseteq Z$

(v) $\vdash x_{\mathbf{A}} \in U_{\mathbf{A}}$

(vi) $\vdash \neg\, (x \in \varnothing_{\mathbf{A}})$

(vii) $\vdash X \in PY \Leftrightarrow X \subseteq Y$

(viii) $\vdash X \subseteq \cap U \Leftrightarrow \forall u \in U \,.\, X \subseteq u$

(ix) $\vdash \cup U \subseteq X \Leftrightarrow \forall u \in U \,.\, u \subseteq X$

(x) $\vdash x \in \{y\} \Leftrightarrow x = y$

(xi) $\vdash \alpha \Rightarrow \tau \in \{\tau : \alpha\}$

Here (i) is the *Axiom of Extensionality,* (iv) the *Axiom of Binary Union,* (vi) the *Axiom of the Empty Set,* (vii) the *Power Set axiom,* (ix) the *Union Axiom* and (x) the *Axiom of Singletons.* These, together with the comprehension axiom, form the core axioms for set theory in $\mathscr{L}$. The set theory is *local* because some of the set theoretic operations, e.g., intersection and union, may be performed only on sets of the same type, that is, "locally". Moreover, variables are constrained to range only over given types—locally—in contrast with the situation in classical set theory where they are permitted to range globally over an all-embracing universe of discourse.

Now define the relation $\sim_S$ on the collection of all $\mathscr{L}$-sets by

$$X \sim_S Y \;\equiv\; \vdash_S X = Y.$$

This is an equivalence relation. An *S-set* is an equivalence class $[X]_S$ —which we normally identify with X—of $\mathscr{L}$-sets under the relation $\sim_S$. An *S-map* f: $X \to Y$ or $X \xrightarrow{f} Y$ is a triple (f, X, Y)—normally identified with f—of S-sets such that $\vdash_S f \in Fun(X, Y)$. X and Y are, respectively, the *domain* dom(f) and the *codomain* cod(f) of f.

Now suppose we are given a term τ such that

$$<x_1, ..., x_n> \in X \vdash_S \tau \in Y.$$

We write $<x_1, ..., x_n> \mapsto \tau$ or simply $\boldsymbol{x} \mapsto \tau$ for

$$\{<<x_1, ..., x_n>, \tau>: <x_1, ..., x_n> \in X\}.$$

If $x_1, ..., x_n$ includes all the free variables of τ and X, Y are S-sets, then $<x_1, ..., x_n> \mapsto \tau$ is an S-map $X \to Y$, which we denote by τ: $X \to Y$ or $X \xrightarrow{\tau} Y$. If $\mathbf{f}$ is a function symbol, we write f for $x \mapsto \mathbf{f}(x)$.

BASIC PROPERTIES OF TOPOSES

As defined in Appendix II, a topos is a category which possesses a terminal object 1, products, a truth-value object Ω, and power objects. It can be shown that every topos is cartesian closed, finitely complete, and has coproducts of arbitrary pairs of its objects.

Given a topos $\mathscr{E}$, and an $\mathscr{E}$-arrow $u\colon A \to \Omega$, we choose $\bar{u}\colon B \to A$ so that

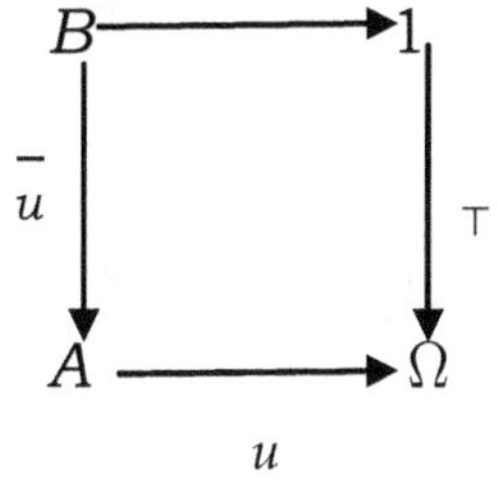

is a pullback and $\overline{\chi(1_A)} = 1_A$.. Note that then $\chi(\bar{u}) = u$.

Now given monics *m, n* with common codomain *A,* write $m \subseteq n$ if there is a commutative diagram of the form

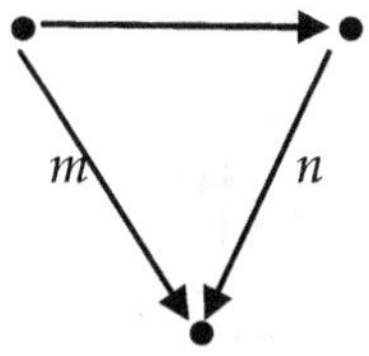

Write $m \sim n$ if $m \subseteq n$ and $n \subseteq m$. Then $\sim$ is an equivalence relation and $m \sim n$ iff there is an isomorphism such that

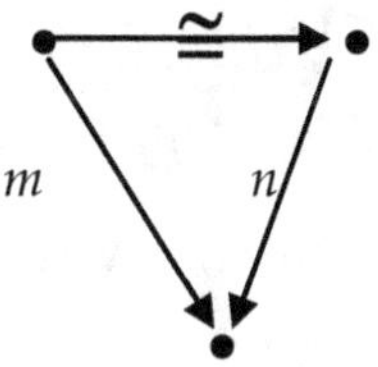

commutes. Equivalence classes under $\sim$ are called *subobjects* of *A*. Write $[m]$ for the equivalence class of *m*. For $u\colon A \to \Omega$, $[\bar{u}]$ is called the subobject of *A* *classified* by *u*. We define $[m] \subseteq [n] \equiv m \subseteq n$. The relation $\subseteq$ – *inclusion* – is a partial ordering on the collection **Sub**(A) of subobjects of A. It is easily shown that $[m] = [n] \equiv \chi(m) = \chi(n)$, so we get a bijection between **Sub**(A) and $\mathscr{E}(A, \Omega)$, the collection of $\mathscr{E}$-arrows $A \rightarrow \Omega$. Define, for $u, v \in \mathscr{E}(A, \Omega)$, $u \leqslant v \equiv \bar{u} \subseteq \bar{v}$. This transfers the partial ordering $\subseteq$ on **Sub**(A) to $\mathscr{E}(A, \Omega)$.

It can be shown by an elementary argument that, in a topos, any diagram of the form

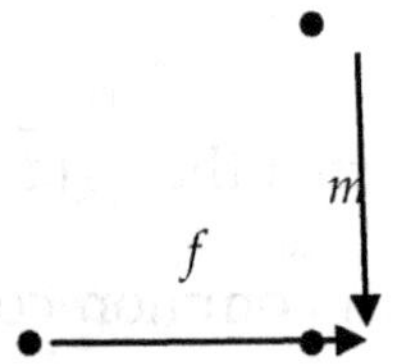

with m monic can be completed to a pullback

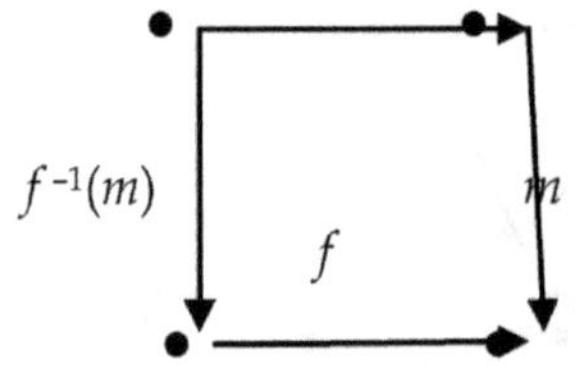

The arrow $f^{-1}(m)$ is called the *inverse image* of m under f. We may in fact take $f^{-1}(m)$ to be $\overline{\chi(m) \circ f}$.

Now define $\delta_A = \langle 1_A, 1_A \rangle: A \rightarrowtail A \times A$, $eq_A = \chi(\delta_A)$, $T_A = \chi(1_A)$. Then $\overline{T_A} = 1_A$, so $u \leqslant T_A$ for all $u \in \mathscr{E}(A, \Omega)$.

Given a pair of monics m, n with common codomain A, we obtain their *intersection* $m \cap n$ by first forming the pullback

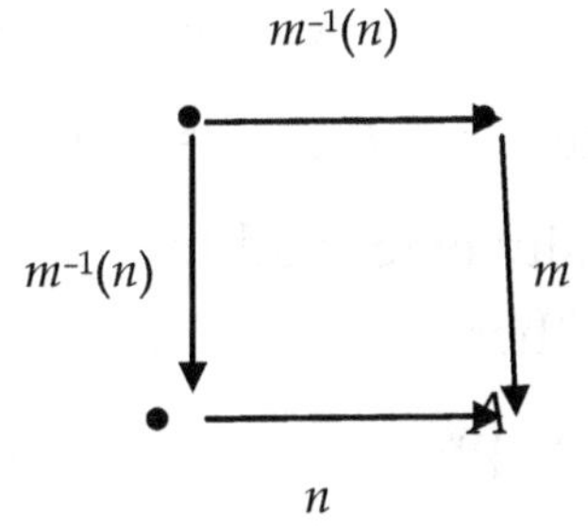

and then defining $m \cap n = n \circ m^{-1}(n) = m \circ n^{-1}(m)$. This turns (**Sub**($A$), $\subseteq$) into a *lower semilattice,* that is, a partially ordered set with meets. We transfer $\cap$ to $\mathscr{E}(A, \Omega)$ by defining $u \wedge v = \chi(\bar{u} \wedge \bar{v})$. This has the effect of turning $\mathscr{E}(A, \Omega)$ into a lower semilattice as well.

TOPOSES AS MODELS OF LOCAL SET THEORIES

Toposes constitute the natural *models* of local set theories in that the latter have been designed to be interpretable in the former. Let $\mathscr{L}$ be a local language and $\mathscr{E}$ a topos. A (*topos-theoretic*) *interpretation I* of $\mathscr{L}$ in $\mathscr{E}$ is an assignment:

- to each type **A**, of an $\mathscr{E}$-object $\mathbf{A}_I$ such that:

  $(\mathbf{A}_1 \times ... \times \mathbf{A}_n)_I = (\mathbf{A}_1)_I \times ... \times (\mathbf{A}_n)_I$,

  $(\mathbf{PA})_I = P\mathbf{A}_I$,

  $\mathbf{1}_I = 1$, the terminal object of $\mathscr{E}$,

  $\Omega_I = \Omega$, the truth-value object of $\mathscr{E}$.

- to each function symbol f: $\mathbf{A} \to \mathbf{B}$, an $\mathscr{E}$-arrow f_I: $\mathbf{A}_I \to \mathbf{B}_I$.

An interpretation I can then be extended to terms of $\mathscr{L}$ in such a way as to yield, for each term τ : **B**, with variables $\mathbf{x} = (x_1, ..., x_n)$, an $\mathscr{E}$-arrow

$$[\![\tau]\!]_{I\mathbf{x}} = [\![\tau]\!]_{\mathbf{x}}: A_1 \times ... \times A_n \to B$$

When **B** is **Ω**, τ is a proposition α, and we can say that α is *true* under *I* if $[\![\alpha]\!]_x = T$[1] This is written $\models_I \alpha$ or $\models_{\mathscr{C}} \alpha$, and we say that α is *I-valid.* This idea is easily extended to the *I*-validity of a sequent Γ : α. If $\Gamma = \{\gamma_1, ..., \gamma_n\}$, then we say that Γ : α is *I*-valid, written $\Gamma \models_I \alpha$ or $\Gamma \models_{\mathscr{C}} \alpha$, if the proposition $(\gamma_1 \wedge ... \wedge \gamma_n) \Rightarrow \alpha$ is *I*-valid. *I* is a *model* of a local set theory *S* if every sequent in *S* is *I*-valid. If $\Gamma \models_I \alpha$ for every model *I* of *S*, we write $\Gamma \models_S \alpha$; if this is the case, we say that α is an *S-consequence of* Γ. Just as for first-order logic, a *completeness theorem* can then be proved in the form

$$\Gamma \models_S \alpha \Leftrightarrow \Gamma \vdash_S \alpha .$$

For any local set theory *S*, the collection of all *S*-sets and maps forms a category $\mathscr{C}(S)$, the *category of S-sets.* This category is actually a topos, a fact proved just as for $\mathcal{S}et$, only arguing formally in *S*.

Categorical properties of objects and arrows in $\mathscr{C}(S)$ are naturally correlated with formal "set-theoretic" properties of the corresponding entities in *S*. Here is a brief table:

$\mathscr{C}(S)$	*S*
$f: X \to Y$ is a monic arrow	$<x,y> \in f,\ <x',y> \in f \vdash_S x = x'$
$f: X \to Y$ is an epic arrow	$y \in Y \vdash_S \exists x.\ <x,y> \in f$
The commutative diagram 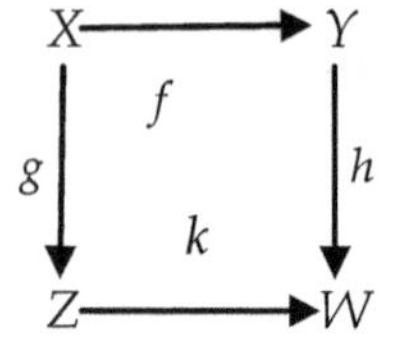 is a pullback.	$y \in Y,\ z \in Z \vdash_S \exists u(<y,u> \in h \wedge <z, u> \in k) \Rightarrow \exists! x(<x, y> \in f \wedge <x, z> \in g)$

[1] See Appendix II for the definition of *T* and other category-theoretic notions.

A topos of the form $\mathscr{C}(S)$ is called a *linguistic* topos. It can be shown that every topos is equivalent to a linguistic one: more precisely, given a topos $\mathscr{E}$, one can produce a local language $\mathscr{L}(\mathscr{E})$ —called the *internal language of* $\mathscr{E}$— and a theory $Th(\mathscr{E})$ in $\mathscr{L}(\mathscr{E})$ for which an equivalence $\mathscr{E} \simeq \mathscr{C}(Th(\mathscr{E}))$ can be established. The language $\mathscr{L}(\mathscr{E})$ has ground type symbols matching the objects of $\mathscr{E}$ other than its terminal and truth-value objects, that is, for each $\mathscr{E}$-object A (other than 1, Ω) we assume given a ground type $\mathbf{A}$ in $\mathscr{L}(\mathscr{E})$. Next, we define for each type symbol $\mathbf{A}$ an $\mathscr{E}$-object $\mathbf{A}_{\mathscr{E}}$ by

$\mathbf{A}_E = \mathbf{A}$ for ground types $\mathbf{A}$,

$(\mathbf{A} \times \mathbf{B})_{\mathscr{E}} = \mathbf{A}_{\mathscr{E}} \times \mathbf{B}_{\mathscr{E}}$ [1]

$(\mathbf{PA})_{\mathscr{E}} = P(\mathbf{A})_{\mathscr{E}}$.

The function symbols of $\mathscr{L}(\mathscr{E})$ are then taken to be triples $(f, \mathbf{A}, \mathbf{B}) = f$ with $f\colon \mathbf{A}_{\mathscr{E}} \to \mathbf{B}_{\mathscr{E}}$ in $\mathscr{E}$. The signature of f is $\mathbf{A} \to \mathbf{B}$.[2]

There is a *natural interpretation*—denoted by $\mathscr{E}$—of $\mathscr{L}(\mathscr{E})$ in $\mathscr{E}$. It is determined by the assignments:

$\mathbf{A}_{\mathscr{E}} = \mathbf{A}$ for each ground type $\mathbf{A}$ $\qquad$ $(f, \mathbf{A}, \mathbf{B})_{\mathscr{E}} = f$.

The local set theory $Th(\mathscr{E})$ is the theory in $\mathscr{L}(\mathscr{E})$ generated by the collection of all sequents $\Gamma : \alpha$ such that $\Gamma \vDash_{\mathscr{E}} \alpha$ under the natural interpretation of $\mathscr{L}(\mathscr{E})$ in $\mathscr{E}$. It can then be shown that

[1] Note that, if we write C for $A \times B$, then while $\mathbf{C}$ is a ground type, $\mathbf{A} \times \mathbf{B}$ is a product type. Nevertheless $\mathbf{C}_{\mathscr{E}} = (\mathbf{A} \times \mathbf{B})_{\mathscr{E}}$.

[2] Note the following: if $f\colon A \times B \to D$, in E, then, writing C for $A \times B$ as in the footnote above, $(f, \mathbf{C}, \mathbf{D})$ and $(f, \mathbf{A} \times \mathbf{B}, \mathbf{D})$ are both function symbols of $\mathscr{L}(\mathscr{E})$ associated with f. But the former has signature $\mathbf{C} \to \mathbf{D}$, while the latter has the different signature $\mathbf{A} \times \mathbf{B} \to \mathbf{D}$.

$$\Gamma \vdash_{Th(\mathcal{E})} \alpha \quad \Leftrightarrow \quad \Gamma \vDash_{\mathcal{E}} \alpha \ .$$

Finally, the canonical functor $F: \mathcal{E} \to \mathcal{C}(Th(\mathcal{E}))$ defined by

$FA = U_{\mathbf{A}}$ *for each* $\mathcal{E}$*-object* A

$Ff = (x \mapsto f(x)\text{): } U_{\mathbf{A}} \to U_{\mathbf{B}}$ *for each* $\mathcal{E}$*-arrow* $f: A \to B$

is an equivalence of categories. This is known as the **Equivalence Theorem.**

A useful consequence of the Equivalence Theorem is that any fact concerning a linguistic topos established by arguing "set-theoretically" within the corresponding local set theory automatically extends to arbitrary toposes.

A local set theory S in a language $\mathcal{L}$ is said to be *well-termed* if:

- whenever $\vdash_S \exists! x\alpha$, there is a term τ of $\mathcal{L}$ whose free variables are those of α with x deleted such that $\vdash_S \alpha(x/\tau)$,

and *well-typed* if

- for any S-set X there is a type symbol $\mathbf{A}$ of $\mathcal{L}$ such that $U_{\mathbf{A}} \cong X$ in $\mathcal{C}(S)$.

A local set theory which is both well-termed and well-typed is said to be *well-endowed.* It can be shown that, for any topos $\mathcal{E}$, $Th(\mathcal{E})$ is well-endowed.

The property of being well-endowed can also be expressed category-theoretically. For a local set theory S, let $\mathcal{T}(S)$—the *category of S-types and terms* —be the subcategory of $\mathcal{C}(S)$ whose objects are all S-sets of the form $U_{\mathbf{A}}$ and whose arrows are all S-maps of the form $x \mapsto \tau$. Then S is well-endowed exactly when the insertion functor $\mathcal{T}(S) \to \mathcal{C}(S)$ is an equivalence of categories.

We shall require the process of *adjoining a generic element to an S-set.* Let X be an S-set of type **PA.** Write $\mathscr{L}(\boldsymbol{c})$ for the language obtained from $\mathscr{L}$ by adding a new function symbol $\boldsymbol{c}: \mathbf{1} \rightarrow \mathbf{A}$ and write c for $\boldsymbol{c}(\star)$. Now let $S(X)$ be the theory in $\mathscr{L}(\boldsymbol{c})$ generated by S together with all sequents of the form $: \alpha(c)$ where $\alpha(x)$ is any formula satisfying $x \in X \vdash_S \alpha(x)$. Clearly $\vdash_{S(X)} c \in X$. It can also be shown that for any formula $\beta(x)$,

$$\vdash_{S(X)} \beta(c) \equiv \vdash_S \forall x{\in}X\ \beta(x).$$

Accordingly, in $S(X)$, c behaves as a *generic* element of X in the sense that, if c has a given property, then *every* element of X has it (and conversely).

THE STRUCTURE OF Ω AND SUB(A)

Let S be a local set theory. We define the *entailment relation* on $\Omega = U_\Omega$ to be the S-set

$$\blacktriangleright = \{<\omega, \omega'>: \omega \Rightarrow \omega'\}.$$

Given an S-set X, we define the *inclusion relation* on PX to be the S-set

$$\preceq_X = \{<u, v> \in PX \times PX: u \subseteq v\}.$$

It follows from facts concerning $\Rightarrow, \wedge, \vee$ already established that

$\vdash_S <\Omega, \blacktriangleright>$ *is a Heyting algebra with top element* $\top$ *and bottom element* $\bot$.

Similarly,

$\vdash_S$ $<PX, \preceq_X>$ *is a Heyting algebra with top element* X *and bottom element* $\varnothing$.

Let Sent(S) be the collection of *sentences* (closed formulas) of $\mathscr{L}$, where we identify two sentences α, β whenever $\vdash_S \alpha \Leftrightarrow \beta$. Define the relation $\leqslant$ on Sent(S) by

$$\alpha \leqslant \beta \equiv \vdash_S \alpha \Rightarrow \beta.$$

Then <Sent(S), $\leqslant$> is a Heyting algebra, called the (external) *algebra of truth values of S*. Its top element is $\top$ and its bottom element $\bot$.

If X is an S-set, write Pow(X) for the collection of all S-sets U such that $\vdash_S U \subseteq V$ and define the relation $\sqsubseteq$ on Pow(X) by $U \sqsubseteq V \equiv \vdash_S U \subseteq V$. Then (Pow($X$), $\sqsubseteq$) is a Heyting algebra, called the (external) *algebra of subsets of X*.

Given a topos $\mathscr{E}$, we can apply all this to the theory $Th(\mathscr{E})$; invoking the fact that $\vdash_{Th(\mathscr{E})} \alpha \equiv \vDash_{\mathscr{E}} \alpha$ then gives

$\vDash_{\mathscr{E}}$ *<Ω, ≼> and <PA, ≼$_A$> are Heyting algebras,*

where A is any $\mathscr{E}$-object. These facts are sometimes expressed by saying that Ω and PA are *internal* Heyting algebras in $\mathscr{E}$.

What are the "internal" logical operations on Ω in $\mathscr{E}$? That is, which arrows $\underline{\wedge}$, $\underline{\vee}$, $\underline{\neg}$, $\underline{\Rightarrow}$ represent $\wedge$, $\vee$, $\neg$, $\Rightarrow$? Working in a linguistic topos and then transferring the result to an arbitrary topos via the Equivalence Theorem shows that, in $\mathscr{E}$,

$\underline{\wedge}: \Omega \times \Omega \to \Omega$ is the characteristic arrow of the monic

$<\top, \top>: 1 \to \Omega \times \Omega$

$\underline{\vee}: \Omega \times \Omega \to \Omega$ is the characteristic arrow of the image of

$$\Omega + \Omega \xrightarrow{<T_\Omega, 1_\Omega> + <1_\Omega, T_\Omega>} \Omega \times \Omega$$

$\underline{\neg}: \Omega \to \Omega$ is the characteristic arrow of $\bot: 1 \to \Omega$.

$\underline{\Rightarrow}: \Omega \times \Omega \to \Omega$ is the characteristic arrow of the equalizer of the pair of arrows $\pi_1, \wedge: \Omega \times \Omega \to \Omega$. (Here we recall that the *equalizer* of a pair of arrows with a common domain is the largest subobject of the domain on which they both agree.)

It can then be shown that these "logical arrows" are the natural interpretations of the logical operations in any topos $\mathscr{E}$, in the sense that, for any interpretation of a language $\mathscr{L}$ in $\mathscr{E}$,

$$[\![\alpha \wedge \beta]\!]_x = \underline{\wedge} \circ [\![<\alpha, \beta>]\!]_x$$

$$[\![\alpha \vee \beta]\!]_x = \underline{\vee} \circ [\![<\alpha, \beta>]\!]_x$$

$$[\![\neg\alpha]\!]_x = \underline{\neg} \circ [\![\alpha]\!]_x$$

$$[\![\alpha \Rightarrow \beta]\!]_x = \underline{\Rightarrow} \circ [\![<\alpha, \beta>]\!]_x$$

We now turn to the "external" formulation of these ideas. First, for any topos $\mathscr{E}$ and any $\mathscr{E}$-object A, (**Sub**(A), $\subseteq$) is a Heyting algebra. For when $\mathscr{E}$ is of the form $\mathscr{C}(S)$, and A an S-set X, we have a natural isomorphism ($Pow(X)$, $\sqsubseteq$) $\cong$ (**Sub**(X), $\subseteq$) given by

$$U \mapsto [(x \mapsto x){:}U \rightarrowtail X]$$

for $U \in Pow(X)$. Since we already know that ($Pow(X)$, $\sqsubseteq$) is a Heyting algebra, so is (**Sub**(X), $\subseteq$). Thus the result holds in any linguistic topos, and hence in any topos.

Since **Sub**(A) $\cong \mathscr{E}(1, PA)$, it follows that $\mathscr{E}(1, PA)$ (with the induced ordering) is a Heyting algebra. And since ($\mathscr{E}(A, \Omega)$, $\leqslant$) $\cong$ (**Sub**(A), $\subseteq$), it follows that the former is a Heyting algebra as well. Taking $A = 1$, we see that the ordered set $\mathscr{E}(1, \Omega)$ of $\mathscr{E}$-elements of Ω is also a Heyting algebra.

Recall that a partially ordered set is *complete* if every subset has a supremum (join) and an infimum (meet). We claim that, for any local set theory S, and any S-set X,

$$\vdash_S \langle \Omega, \preceq \rangle \textit{ and } \langle PX, \subseteq \rangle \textit{ are complete.}$$

For we have

$$u \subseteq \Omega \vdash_S (\top \in u) \text{ is the } \preceq\text{-join of } u,$$

$$u \subseteq \Omega \vdash_S (\forall \omega \in u.\ \omega) \text{ is the } \preceq\text{-inf of } u,$$

$$v \subseteq X \vdash_S \bigcup v \text{ is the } \subseteq\text{-join of } v,$$

$$v \subseteq X \vdash_S \bigcap v \text{ is the } \subseteq\text{-meet of } v.$$

To prove, e.g., the first assertion, observe that, first,

$$u \subseteq \Omega, \omega \in u, \omega \vdash_S \omega \in u \wedge \omega = \top \vdash_S \top \in u,$$

so

$$u \subseteq \Omega, \omega \in u \vdash_S \omega \Rightarrow (\top \in u) \wedge \omega = \top \vdash_S \omega \preceq (\top \in u)$$

whence

$$u \subseteq \Omega \vdash_S \omega \in u \Rightarrow \omega \preceq (\top \in u),$$

and thus

$$u \subseteq \Omega \vdash_S \top \in u \textit{ is an } \preceq\textit{-upper bound for } u.$$

Also

$$u \subseteq \Omega, \forall \omega \in u(\omega \Rightarrow \alpha), (\top \in u) \vdash_S \top \Rightarrow \alpha \vdash_S \alpha,$$

whence

$$u \subseteq \Omega, \forall \omega \in u(\omega \Rightarrow \alpha), (\top \in u) \vdash_S \alpha,$$

i.e.,

$$u \subseteq \Omega,, \alpha \textit{ is an } \preceq\textit{-upper bound for } u \vdash_S (\top \in u) \preceq \alpha,$$

which establishes the first assertion.

As a consequence, for any topos $\mathscr{E}$,

$$\vDash_{\mathscr{E}} \langle \Omega, \preceq \rangle \textit{ and } \langle PA, \subseteq \rangle \textit{ are complete.}$$

That is, Ω and PA are *internally complete* in $\mathscr{E}$.

EXAMPLES OF TOPOSES

One of F. W. Lawvere's most penetrating insights[1] was to conceive of a topos as a universe of *variable sets*. Here are some examples.

To begin with, consider the category $\mathcal{B}iv$ of bivariant sets introduced above. This is a topos in which the truth value object Ω in has 3 (rather than 2) elements. For if (m, X) is a subobject of Y in Biv, then we may take $X_0 \subseteq Y_0$, $X_1 \subseteq Y_1$, f_0 and f_1 identity maps, and p to be the restriction of q to X_0. Then for any $y \in Y$ there are three possibilities, as depicted below: (0) $y \in X_0$, (1) $q(y) \in X_1$ and $y \notin X_0$, and (2) $q(y) \notin X_1$.

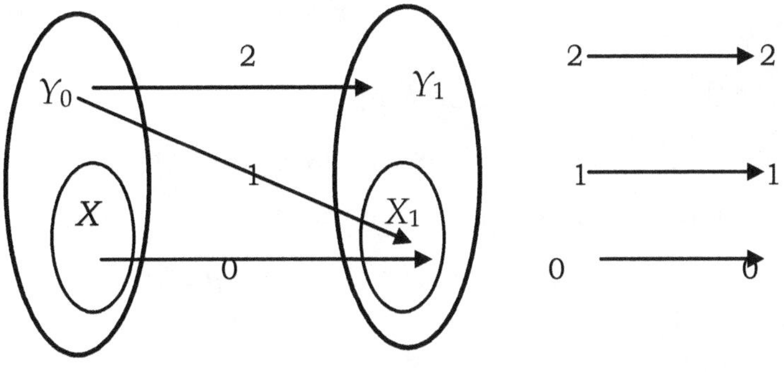

So if 2 = {0, 1} and 3 = {0, 1, 2} we take Ω to be the bivariant set $3 \to 3$ with $0 \mapsto 1, 1 \mapsto 1, 2 \mapsto 2$.

More generally, we may consider sets varying over n, or ω, or any totally ordered set of stages. Objects in $\mathcal{S}et^n$ are "sets through n successive stages", that is, $(n-1)$-tuples of maps

$$X_0 \xrightarrow{f_0} X_1 \xrightarrow{f_1} X_2 \xrightarrow{f_2} \ldots X_{n-2} \xrightarrow{f_{n-2}} X_{n-1}.$$

[1] See, e.g., Lawvere [1972], [1976].

Objects in $\mathcal{S}et^{\omega}$ are "sets through discrete time", that is, infinite sequences of maps

$$X_0 \xrightarrow{f_0} X_1 \xrightarrow{f_1} X_2 \xrightarrow{f_2} \ldots.$$

Still more generally, we may consider the category $\mathcal{S}et^{P}$ of sets varying over a poset P. As objects this category has *functors*[1] $P \to \mathcal{S}et$, i.e., maps F which assign to each $p \in P$ a set $F(p)$ and to each $p, q \in P$ such that $p \leqslant q$ a map F_{pq}: $F(p) \to F(q)$ satisfying:

$p \leqslant q \leqslant r$ implies that 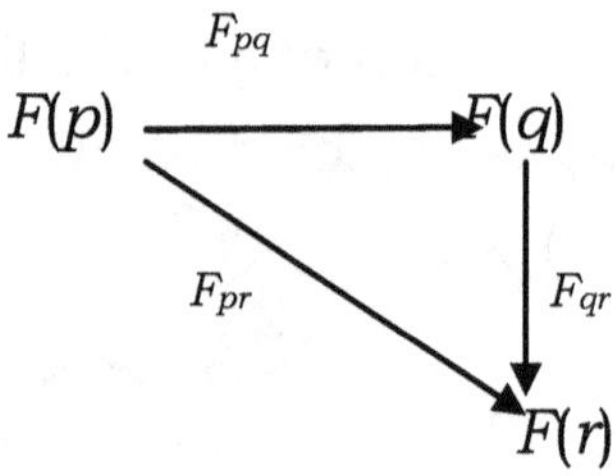

commutes

and F_{pp} is the identity map on $F(p)$.

An *arrow* η: $F \to G$ in $\mathcal{S}et^{P}$ is a *natural transformation* between F and G, which in this case is an assignment of a map η_p: $F(p) \to G(p)$ to each $p \in P$ in such a way that, whenever $p \leqslant q$, the diagram

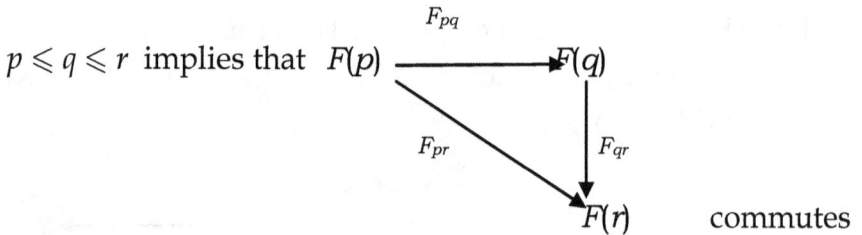

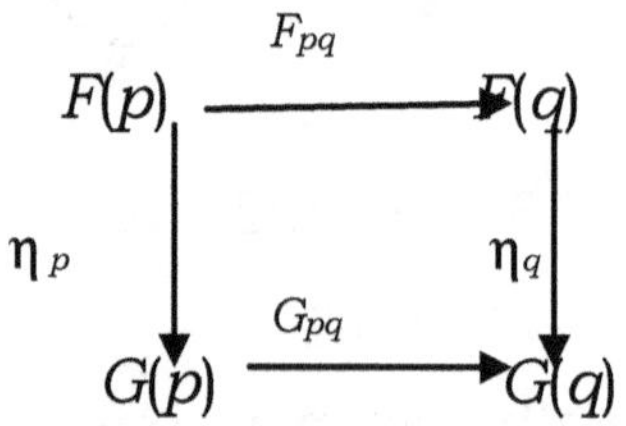

commutes.

To determine Ω in $\mathcal{S}et^{P}$ we define a (pre)*filter over* $p \in P$ to be a subset U of $O_p = \{q \in P : p \leqslant q\}$ such that $q \in U, r \geq q \Rightarrow r \in U$. Then

[1] Recall that any preordered set, and in particular any poset, may be regarded as a category.

$$\Omega(p) = \text{set of all filters over } p,$$

$$\Omega_{pq}(U) = U \cap O_q \ \text{ for } p \leqslant q,\ U \in \Omega(p).$$

The *terminal object* 1 in $\mathcal{S}et^{\mathbf{P}}$ is the functor on P with constant value $1 = \{0\}$ and $t: 1 \to \Omega$ has $t_p(0) = O_p$ for each $p \in P$.

Objects in $\mathcal{S}et^{\mathbf{P}^*}$ —where P^* is the poset obtained by reversing the order on P—are called *presheaves* on P. In particular, when P is the partially ordered set $\mathcal{O}(X)$ of open sets in a topological space X, objects in $\mathcal{S}et^{\mathcal{O}(X)}$ called *presheaves on* X. So a presheaf on X is an assignment to each $U \in \mathcal{O}(X)$ of a set $F(U)$ and to each pair of open sets U, V such that $V \subseteq U$ of a map $F_{UV} : F(U) \to F(V)$ such that, whenever $W \subseteq U \subseteq V$, the diagram

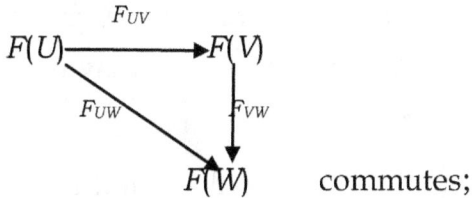

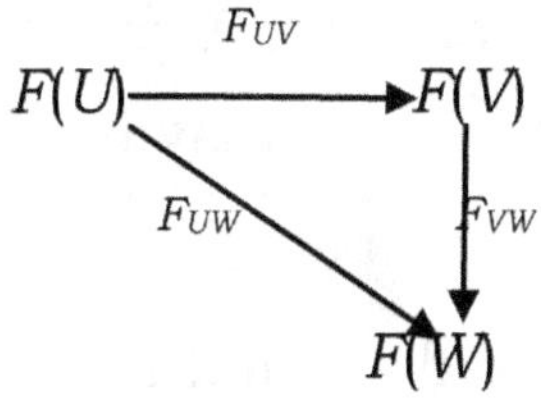

commutes;

and F_{UU} is the identity map on $F(U)$.

If $s \in F(U)$, write $s|_V$ for $F_{UV}(s)$—the *restriction* of s to V. A presheaf F is a *sheaf* if whenever $U = \bigcup_{i \in I} U_i$ and we are given a set $\{s_i: i \in I\}$ such that $s_i \in F(U_i)$ for all $i \in I$ and $s_i|_{U_i \cap U_j} = s_j|_{U_i \cap U_j}$ for all $i, j \in I$, then there is a *unique* $s \in F(U)$ such that $s|_{U_i} = s_i$ for all $i \in I$. For example, $C(U)$ = set of continuous real-valued functions on U, and $s|_V$ = restriction of s to V defines the sheaf of continuous real-valued functions on X.

It can be shown that the category $\mathcal{S}hv(X)$ of sheaves on X (that is, the full subcategory of $\mathcal{S}et^{\mathbf{P}^*}$ whose objects are sheaves) is a topos which is equivalent as a category to the category $\mathcal{E}sp(X)$

of displayed spaces on X introduced above. It follows that the latter is also a topos.

The idea of a set varying over a poset can be naturally extended to that of a set varying over an arbitrary *small category.* Given a small category $\mathscr{C}$, we introduce the category $\mathcal{S}et^{\mathscr{C}}$ of *sets varying over* $\mathscr{C}$. Its objects are all functors $\mathscr{C} \to \mathcal{S}et$, and its arrows all natural transformations between such functors. Again, it can be shown that $\mathcal{S}et^{\mathscr{C}}$ is a topos.

An important special case arises when $\mathscr{C}$ is a one-object category, that is, a *monoid.* To be precise, a monoid is a pair $\mathbf{M} = (M, \cdot)$ with M a set and $\cdot$ a binary operation on M satisfying the associative law $\alpha \cdot (\beta \cdot \gamma) = (\alpha \cdot \beta) \cdot \gamma$ and possessing an identity element 1 satisfying $1 \cdot \alpha = \alpha \cdot 1 = \alpha$. (Note that a *group* is just a monoid with inverses, that is, for each α there is β for which $\alpha \cdot \beta = \beta \cdot \alpha = 1$.) Any object in $\mathcal{S}et^{\mathbf{M}}$ may be identified with a *set acted on by* **M**, or **M**-*set,* that is, a pair $(X, \bullet)$ with $\bullet$ a map $M \times X \to X$ satisfying $(\alpha \cdot \beta) \bullet x = \alpha \bullet (\beta \bullet x)$ and $1 \bullet x = x$. An arrow f: $(X, \bullet) \to (Y, \bullet)$ is an *equivariant* map f: $X \to Y$, i.e, such that $f(\alpha \bullet x) = \alpha \bullet f(x)$. The subobject classifier Ω in $\mathcal{S}et^{\mathbf{M}}$ is the collection of all *left ideals* of **M**, i.e. those $I \subseteq M$ for which $\alpha \in I, \beta \in M \Rightarrow \beta \cdot \alpha \in I$. The action of M on Ω is *division,* viz. $\alpha \bullet I = \{\beta \in M: \beta \cdot \alpha \in I\}$[1]. The truth arrow t: $1 \to \Omega$ is the map with value M.

Toposes can also arise as categories of "sets with a generalized equality relation", with arrows preserving that relation in an appropriate sense. Some of the most important

[1] This is because if X is a sub-**M**-set of Y, each $y \in Y$ is naturally classified by the left ideal $\{\alpha \in M: \alpha \bullet y \in X\}$.

examples in this regard are the *categories of Heyting algebra-valued sets*. Given a complete Heyting algebra H, an *H-valued set* is a pair (I, δ) consisting of a set I and a map $\delta: I \times I \to H$ (the "generalized equality relation" on I) satisfying the following conditions, in which we write $\delta_{ii'}$ for $\delta(i, i')$ (and similarly below):

$$\delta_{ii'} = \delta_{i'i} \quad \text{(symmetry)}$$

$$\delta_{ii'} \wedge \delta_{i'i''} \leqslant \delta_{ii''} \text{ (substitutivity)}$$

The category $\mathcal{S}et_H$ of H-valued sets has as objects all H-valued sets. A $\mathcal{S}et_H$-arrow $f: (I, \delta) \to (J, \varepsilon)$ is a map $f: I \times J \to H$ such that

$$\delta_{ii'} \wedge f_{ij} \leqslant f_{i'j} \qquad f_{ij} \wedge \varepsilon_{jj'} \leqslant f_{ij'} \quad \text{(preservation of identity)}$$

$$f_{ij} \wedge f_{ij'} \leqslant \varepsilon_{jj'} \quad \text{(single-valuedness)}$$

$$\bigvee_{j \in J} f_{ij} = \delta_{ii} \quad \text{(defined on } I\text{)}$$

The composite $g \circ f$ of two arrows $f: (I, \delta) \to (J, \varepsilon)$ and $g: (J, \varepsilon) \to (K, \eta)$ is given by

$$(g \circ f)_{ik} = \bigvee_{j \in J} f_{ij} \wedge g_{jk}.$$

Then $\mathcal{S}et_H$ is a topos in which the subobject classifier is the H-valued set $(H, \blacklozenge)$, where $\blacklozenge$ is the equivalence operation on H.

It can be shown that, for any topological space X, $\mathcal{S}et_{\mathcal{O}(X)}$ is equivalent to $\mathcal{S}hv(X)$ and so also to $\mathcal{E}sp(X)$; and, for any complete Boolean algebra B, $\mathcal{S}et_B$ is equivalent to the category $\mathcal{F}uz_B$ of B-fuzzy sets.

THE CHOICE RULE AND OTHER PRINCIPLES IN LOCAL SET THEORIES

Let S be a local set theory in a language $\mathscr{L}$. We make the following definitions.

- S is *classical* if $\vdash_S \forall\omega(\omega \vee \neg\omega)$. This is the full Law of Excluded Middle for S.
- S is *sententially classical* if $\vdash_S \sigma \vee \neg\sigma$ for any sentence σ. This is the Law of Excluded Middle for sentences.
- S is *complete* if $\vdash_S \sigma$ or $\vdash_S \neg\sigma$ for any sentence σ.
- For each S-set A : **PB** let $\Delta(A)$ be the set of closed terms τ such that $\vdash_S \tau \in A$. A is *standard* if for any formula α with at most the variable x : **B** free the following rule is valid:

  $$\frac{\vdash_S \alpha(x/\tau) \textit{ for all } \tau \textit{ in } \Delta(A)}{\vdash_S \forall x{\in}A\, \alpha}$$

  S is *standard* if every S-set is so.
- If A is an S-set of type **PB**, an *A-singleton* is a closed term U of type **PB** such that $\vdash_S U \subseteq A$ and $\vdash_S \forall x{\in}U\forall y{\in}U.\ x = y$. X is said to be *near-standard* if for any formula α with at most the variable x : **B** free the following rule is valid:

  $$\frac{\vdash_S \forall x{\in}U\alpha(x) \textit{ for all A-singletons } U}{\vdash_S \forall x{\in}A\, \alpha}$$

  S is *near-standard* if every S-set is so.
- S is *witnessed* if for any type symbol **B** of $\mathscr{L}$ and any formula α with at most the variable x : **B** free the following rule is valid:

$$\frac{\vdash_S \exists x\, \alpha}{\vdash_S \alpha(x/\tau) \;\; \textit{for some closed term } \tau : \mathbf{B}.}$$

- S is *choice* if, for any S-sets X, Y and any formula α with at most the variables x, y free the following rule (the *choice rule*) is valid:

$$\frac{\vdash_S \forall x \in X\, \exists y \in Y\, \alpha(x, y)}{\vdash_S \forall x \in X\, \alpha(x, fx) \;\; \textit{for some } f\!: X \to Y}$$

- S is *internally choice* if under the conditions of the previous definition

$$\forall x \in X\, \exists y \in Y\, \alpha(x, y) \vdash_S \exists f \in Fun(X,Y)$$
$$\forall x \in X\, \exists y \in Y\, [\alpha\,(x, y) \wedge \langle x, y\rangle \in f].$$

- S is *Hilbertian* if for any formula with at most the variables x: **A** and y: **B** free such that $\nvdash_S \neg\exists x \exists y \alpha(x,y)$ there is a term $\tau(x)$: **B** such that $\vdash_S \forall x[\exists y \alpha(x, y) \Rightarrow \alpha(x, \tau(x))]$.[1]

- S is *Zornian* if, for any pair of S-sets E, $\leqslant$, the following rule (the *Zorn* rule) is valid:

$$\frac{\vdash_S \; (E, \leqslant) \textit{ is a strongly inductive partially ordered set}}{\textit{There is an S-element } m \textit{ of } E \textit{ such that } m \textit{ is maximal in } E\textit{, that is, } \vdash_S \forall x \in E\, [m \leq x \Rightarrow m = x]}$$

- An S-set X is *discrete* if

$$\vdash_S \forall x \in X\, \forall y \in X.\; x = y \vee x \neq y.$$

- A *complement* for an S-set X : **PA** is an S-set Y : **PA** such that $\vdash_S X \cup Y = A \;\wedge\; X \cap Y = \varnothing$. An S-set that has a complement is said to be *complemented.*

[1] The term *Hilbertian* is used here because the term $\tau(x)$ here is evidently analogous to the Hilbert ε-term determined by the formula α.

- S is *full* if for each set I there is a type symbol $\mathbf{I}^\wedge$ of the language $\mathscr{L}$ of S together with a collection $\{i^\wedge : i \in I\}$ of closed terms each of type $\mathbf{I}^\wedge$ satisfying the following:

  (i) If $\vdash_S i^\wedge = j^\wedge$ then $i = j$.

  (ii) For any I - indexed family $\{\tau_i : i \in I\}$ of closed terms of common type $\mathbf{A}$, there is a term $\tau(x)$: $\mathbf{A}$, x:$\mathbf{I}^\wedge$ such that $\vdash_S \tau_i = \tau(i^\wedge)$ for all $i \in I$ and, for any term $\sigma(x) : \mathbf{A}, x : \mathbf{I}^\wedge$, if $\vdash_S \tau_i = \sigma(i^\wedge)$ for all $i \in I$, then $\vdash_S \tau = \sigma$.

$\mathbf{I}^\wedge$ may be thought of as the representative in S of the set I.

We now prove the

Generalization Principle for hatted type symbols. *Suppose that S is full. Then the following rule is valid for any formula $\alpha(x)$ with $x : \mathbf{I}^\wedge$*

$$\frac{\vdash_S \alpha(i^\wedge) \quad \textit{for all } i \in I}{\vdash_S \forall x\, \alpha}$$

and similarly for more free variables. In particular, $\mathbf{I}^\wedge$ is standard.

Proof. Assume the premises. Then for any $i \in I$ we have $\vdash_S \alpha(i^\wedge) = \top$ and it follows from the uniqueness condition that $\vdash_S \alpha(x) = \top$, whence $\vdash_S \forall x \alpha$. ■

We next establish some facts concerning these notions. In formulating our arguments we shall assume, with one exception (Proposition 7) that our background metatheory is constructive, in that no use of the metalogical Law of Excluded Middle will be made.

Proposition 1. *Any of the following conditions is equivalent to the classicality of S:*

(i) $\vdash_S \Omega = \{\top, \bot\}$

(ii) $\vdash_S \neg\neg\omega \Rightarrow \omega$

(iii) $\vdash_S$ *Ω is a Boolean algebra*

(iv) *any S-set is complemented,*

(v) *any S-set is discrete,*

(vi) *Ω is discrete,*

(vii) $\vdash_S 2 = \{0, 1\}$ *is well-ordered under the usual ordering,.*

Proof. (iv) If *S* is classical, clearly $\{x: x \notin X\}$ is a complement for *X*. Conversely, if $\{\top\}$ has a complement *U*, then

$$\vdash_S \omega \in U \Rightarrow \neg(\omega = \top) \Rightarrow \neg\omega \Rightarrow \omega = \bot.$$

Hence $\vdash_S U = \{\bot\}$, whence $\vdash_S \Omega = \{\top\} \cup U = \{\top, \bot\}$.

(vi) If Ω is discrete, then $\vdash_S \omega = \top \vee \neg(\omega = \top)$, so $\vdash_S \omega \vee \neg\omega$.

(vii) If *S* is classical, then 2 is trivially well-ordered under the usual well-ordering. Conversely, if 2 is well-ordered, take any formula α, and define $X = \{x \in 2: x = 1 \vee \alpha\}$. Then *X* has a least element, *a*, say. Clearly $\vdash_S a = 0 \Leftrightarrow \alpha$, so, since $\vdash_S a = 0 \vee a = 1$, we get $\vdash_S a = 1 \Leftrightarrow \neg\alpha$, and hence $\vdash_S \alpha \vee \neg\alpha$. ■

Proposition 2. *For well-termed S, S is choice iff S internally choice and witnessed.*

Proof. Suppose *S* is choice. If $\vdash_S \exists x\alpha$, let $u : \mathbf{1}$ and define $\beta(u, x) \equiv \alpha(x)$. Then $\vdash_S \forall u{\in}1\exists x{\in}X\beta(u,x)$. Now choice yields an *S*-map $f: 1 \to X$ such that $\vdash_S \forall u{\in}1\beta(u,f(u))$ i.e., $\vdash_S \beta(\star,f\star)$ or $\vdash_S \alpha(f\star)$. By well-termedness, $f\star$ may be taken to be a closed term τ, and we then have $\vdash_S \alpha(\tau)$. So *S* is witnessed.

To derive internal choiceness from choiceness, we argue as follows: let

$$X^* = \{x \in X: \exists y \in Y \alpha(x, y)\}.$$

Then $\vdash_S \forall x \in X^* \exists y \in Y \alpha(x,y)$. Accordingly choiceness yields a map $f: X^* \to Y$ such that

$$\vdash_S \forall x \in X^* \alpha(x, fx), \text{ i.e. } \vdash_S \forall x \in X^* \exists y \in Y[\langle x, y\rangle \in f \wedge \alpha(x, y)].$$

Now

$$\forall x \in X \exists y \in Y\, \alpha(x, y) \vdash_S X = X^* \vdash_S f \in Fun(X, Y)$$

so

$$\forall x \in X \exists y \in Y\, \alpha(x, y) \vdash_S \forall x \in X \exists y \in Y[\langle x, y\rangle \in f \wedge \alpha(x, y)].$$

Hence

$\forall x \in X\, \exists y \in Y\, \alpha(x, y) \vdash_S$

$$\exists f \in Fun(X,Y)\, \forall x \in X\, \exists y \in Y\, [\alpha\, (x, y) \wedge \langle x, y\rangle \in f],$$

as required. The converse is easy. ■

Proposition 3. *If S is well-endowed, then S is choice iff S(X) is witnessed for every S-set X.*

Proof. Suppose S is choice and $\vdash_{S(X)} \exists y \alpha(y)$. We may assume that X is of the form $U_{\mathbf{A}}$, in which case α is of the form $\beta(x/c, y)$ with $x : \mathbf{A}$. From $\vdash_{S(X)} \exists y \beta(x/c, y)$ we infer $\vdash_S \forall x \exists y \beta(x/c, y)$. So using the choiceness and well-termedness of S we obtain a term $\tau(x)$ such that $\vdash_S \forall x \beta(x, \tau(x))$. Hence $\vdash_{S(X)} \beta(c, \tau(c))$, i.e., $\vdash_{S(X)} \alpha(\tau(c))$. Therefore S_X is witnessed.

Conversely, suppose S_X is witnessed for every S-set X, and that $\vdash_S \forall x \in X \exists y \in Y\, \alpha(x, y)]$. Then $\vdash_{S(X)} \exists y \in Y\, \alpha(c, y)]$, so there is a closed $\mathscr{L}_X$-term τ such that $\vdash_{S(X)} \tau \in Y \wedge \alpha(c, \tau)$. But τ is $\tau'(x/c)$ for some $\mathscr{L}$-term $\tau'(x)$. Thus $\vdash_{S(X)} \tau'(c) \in Y \wedge \alpha(c, \tau'(c))$, whence

$\vdash_S \forall x{\in}X\ [\tau'(x) \in Y \wedge \alpha(x, \tau')]$. Defining $f = (x \mapsto \tau')$: $X \to Y$ then gives $\vdash_S \forall x \in X\ \alpha(x, fx)]$ as required.

Proposition 4 (Diaconescu's Theorem). *If S is choice, then S is classical.*

Proof. Step 1. *If S is choice,* then S_I *is choice for any S-set I.*

Proof of step 1. Suppose that S is choice, and

$$\vdash_{S(I)}\ \forall x \in X(c)\ \exists y \in Y(c)\ \alpha\ (x, y, c).$$

Then

$$\vdash_S\ \forall x \in X(i)\ \exists y \in Y(i)\ \alpha\ (x, y, i).$$

Define

$$X^* = \{<x, i>: x \in X(i) \wedge i \in I\}, \quad Y^* = \bigcup_{i \in I} Y(i),$$

$$\beta(u, i)\ \equiv\ \exists x \in X(i) \exists i \in I[u = <x, i> \wedge \alpha(x, y, i) \wedge y \in Y(i)].$$

Then $\vdash_S\ \forall u \in X^*\ \exists y \in Y^*\ \beta(u, y)$. So choice yields f^*: $X^* \to Y^*$ such that $\vdash_S\ \forall u \in X^*\ \beta(u, f^*u)$, i.e.

$$\vdash_S\ \forall i \in I\ \forall x \in X(i)\ \alpha(x, f^*(<x\ ,i>, i) \wedge f^*(<x\ ,i>) \in Y(i)],$$

whence

$$\vdash_S\ \forall x \in X(c)\ \alpha(x, f^*(<x\ c>, c) \wedge f^*(<x\ ,c>) \in Y(c)],$$

Now define $f = (x \mapsto f^*(<x, c>))$. Then f: $X(c) \to Y(c)$ in S_I and

$$\vdash_{S(I)}\ \forall x \in X(c)\ \alpha(x, fx, c).$$

This completes the proof of step 1.

Step 2. *If S is choice, then S is sententially classical.*

Proof of step 2. Define $2 = \{0, 1\}$ and let $X = \{u \subseteq 2: \exists y.y \in u\}$. Then

$$\vdash_S\ \forall u \in X\ \exists y \in 2\ .\ y \in u.$$

So by choice there is f: $X \to 2$ such that

$$\vdash_S\ \forall u \in X\ .\ fu \in u.$$

Now let σ be any sentence; define

$$U = \{x \in 2: x = 0 \vee \sigma\},\ V = \{x \in 2: x = 1 \vee \sigma\},$$

Then $\vdash_S U \in X \wedge V \in X$, so, writing $a = fU$, $b = fV$, we have

$$\vdash_S [a = 0 \vee \sigma] \wedge [b = 1 \vee \sigma],$$

whence

$$\vdash_S [a = 0 \wedge b = 1] \vee \sigma,$$

so that

(*) $$\vdash_S a \neq b \vee \sigma.$$

But $\sigma \vdash_S U = V \vdash_S a = b$, so that $a \neq b \vdash_S \neg\sigma$. It follows from this and (*) that

$$\vdash_S \sigma \vee \neg\sigma,$$

as claimed. This establishes step 2.

Moral of step 2: if set doubletons have choice functions, then logic is classical.

Step 3. *S is classical iff S(Ω) is sententially classical.* This follows from the fact that, if ϖ is the generic element of Ω introduced in $S(\Omega)$, then $\vdash_S \forall\omega(\omega \vee \neg\omega) \equiv \vdash_{S(\Omega)} \varpi \vee \neg\varpi$.

To complete the proof of Diaconescu's theorem, we now have only to observe that *S is choice* ⇒ S_Ω *is choice* ⇒ S_Ω *is sententially classical* ⇒ *S is classical.* ■

It follows immediately from Diaconescu's theorem that, since not every local set theory is classical, **AC** is *independent* of pure local set theory.

Proposition 5. *If S is well-termed and choice, then S is near-standard.*

Proof. Assume that *S* is choice. To show that *S* is near-standard, we first obtain, for any *S*-set *A* of type **PB** and any formula $\alpha(x)$ with x : **B**, an A-singleton V for which (1) $\vdash_S$ $\forall x \in V \neg\alpha$ and (2) $\vdash_S$ $\exists x \in A \neg\alpha \Rightarrow \exists x.\ x \in V$. Let $X = \{u: \exists x \in A\ \neg\alpha\}$ with u : **1** and $Y = \{x \in A: \alpha\}$. Then $\vdash_S$ $\forall u \in X \exists x \in Y \neg\alpha$, so by choice there is a map f: $X \rightarrow Y$ such that $\vdash_S$ $\forall u \in X \neg\alpha(x/fu)$. If we define $V = \{x: <\star, x> \in f\}$, it is easily checked that V is an A-singleton satisfying conditions (1) and (2).

Now to show that S is near-standard, suppose that $\vdash_S \forall x \in U\ \alpha$ for any A-singleton U. Then in particular $\vdash_S \forall x \in V\ \alpha$, which with (1) gives $\vdash_S \neg\exists x.\ x \in V$. We then deduce, using (2), that $\vdash_S \neg\exists x \in A \neg\alpha$. Since S, being choice, is also classical (Prop. 4), it follows that $\vdash_S \forall x \in A\ \alpha$. Hence S is near-standard. ■

Proposition 6. *If S is well-termed, choice and complete, then S* is *standard.*

Proof. Assume the premises. Then by Prop. 5, S is near-standard. We use completeness to show that S is standard. Suppose then that $\vdash_S \alpha(x/\tau)$ for all $\tau \in \Delta(A)$. If U is an A-singleton, then, assuming S is complete, either $\vdash_S \exists x.\ x \in U$ or $\vdash_S \neg\exists x.\ x \in U$. In the former case, the well-termedness of S yields a closed term τ such that $U = \{\tau\}$ and from $\vdash_S \alpha(x/\tau)$ it then follows that $\vdash_S \forall x \in U\ \alpha$. If, on the other hand, $\vdash_S \neg\exists x.\ x \in U$, then clearly $\vdash_S \forall x \in U\ \alpha$. So $\vdash_S \forall x \in U\ \alpha$ for any A-singleton U, and the near-standardness of S yields $\vdash_S \forall x \in A\ \alpha$, showing that S is standard. ■

Proposition 7. *Suppose that S is well-endowed.* Then

- **(i)** *S is Hilbertian iff S is complete and choice;*
- **(ii)** *if S Hilbertian, then S is standard.*

Proof. (i) Suppose that S is Hilbertian. To show that it is complete, let σ be a sentence and suppose that $\nvdash_S \neg\sigma$. Then $\nvdash_S \neg\exists x\sigma$ with x: **1.** Since S is well-endowed, there is a type **A** and a term $\xi(y)$: **1** with y: **A** such that the map $y \mapsto \xi(y)$ is an isomorphism between $U_{\mathbf{A}}$ and $\{x: \sigma\}$. Since $\nvdash_S \neg\exists x\sigma$, and $U_{\mathbf{A}}$ and $\{x: \sigma\}$ are isomorphic, it follows that $\nvdash_S \neg\exists y.\ y = y$. If S is Hilbertian, there is a then closed term θ: **A** (such that $\vdash_S \theta = \theta$, but this is redundant). That being the case, the closed term $\tau = \xi(\theta)$ satisfies $\vdash_S\ \tau \in \{x: \sigma\}$, and from this it follows immediately that $\vdash_S \sigma$. Using the Law of Excluded Middle in the metatheory, we conclude that S is complete.

We now use the completeness of S in showing that it is choice. Thus suppose $\vdash_S\ \forall x \in X\ \exists y \in Y\ \alpha\ (x, y)$, with X: **PA** and Y: **PB**, and let $\beta(x, y)$ be the formula $\alpha(x, y) \wedge x \in X \wedge y \in Y$. Since S is complete, either (a) $\vdash_S\ \exists x.\ x \in X$ or (b) $\vdash_S\ \neg\exists x.\ x \in X$. In case (a) we have $\vdash_S\ \exists x\exists y\beta(x, y)$, and so since S is Hilbertian there is $\tau(x)$: **B** such that $\vdash_S \forall x[\exists y\ \beta\ (x,\ y) \Rightarrow \beta(x,\ \tau(x)]$. Setting $f = \{<x, y \in X \times Y: y = \tau(x)\}$, we find that $f\colon X \to Y$ and $\vdash_S\ \forall x \in X\ \exists y\ \alpha(x, fx)$. In case (*b*) $X = \varnothing_{\mathbf{A}}$; putting $f = \varnothing_{\mathbf{A}} \times Y$, we again find that $f\colon X \to Y$ and $\vdash_S\ \forall x \in X\ \exists y\ \alpha(x, fx)$. So S is choice.

Conversely, suppose that S is complete and choice. Then by Prop. 4, S is also classical. To show that S is Hilbertian, suppose that $\nvdash_S \neg\exists x\exists y\alpha(x, y)$ with x: **A,** y: **B.** Then by the completeness of S, $\vdash_S \exists x\exists y\alpha(x, y)$. Since S is choice, by Prop. 2 it is witnessed, so there is a closed term θ: **B** such that $\vdash_S\ \exists x\alpha(x,\ \theta)$. Now set

$X = \{x: \exists y\alpha(x,y)\}$ and $Y = \{y: \exists x\alpha(x,y)\}$. Then $\vdash_S \forall x \in X\, \exists y \in Y\, \alpha(x, y)$. Since S is choice, there is $f: X \to Y$ such that $\vdash_S \forall x \in X\, \exists y \in Y\, \alpha(x, f)$. Now let $g = f \cup \{<x, \theta>: x \notin X\}$. From the classicality of S it follows that $f: A \to B$, and so by the well-termedness of S there is a term $\tau(x)$: **B** such that $\vdash_S \tau(x) = gx$. It is now easily verified that $\vdash_S \forall x[\exists y\alpha(x, y) \Rightarrow \alpha(x, \tau(x))]$. Hence S is Hilbertian.

(ii). This follows immediately from **(i)** and Prop. 6. ■

Proposition 8. *Let S be a well-termed, near-standard full local set theory. Then* **MZL** *implies that S is Zornian.*[1]

Proof. Suppose $\vdash_S$ *(E, ≤) is a strongly inductive partially ordered set*. Let E^* be the set of S-elements of E. Since $\vdash_S$ *∅ is a chain in E*, $\bigvee\varnothing \in E^*$ and so E^* is nonempty. Partially order E^* by stipulating that $a \leq^* b$ if $\vdash_S a \leq b$. We show first that $(E^*, \leq^*)$ has a maximal element. To do this we show that $(E^*, \leq^*)$ is strongly inductive. To this end, let $C = \{c_i: i \in I\}$ be a chain in E^*. Since S is full, there is a term $\tau(x)$ such that $\vdash_S \tau(i^\wedge) = c_i$ for all $i \in I$. Since $\vdash_S c_i \leq c_j \vee c_j \leq c_i$. so that $\vdash_S \tau(i^\wedge) \leq \tau(i^\wedge) \vee \tau(j^\wedge) \leq \tau(i^\wedge)$ for every $i, j \in I$, it follows from the Generalization Principle that $\vdash_S \forall x \forall y[\tau(x) \leq \tau(y) \vee \tau(y) \leq \tau(x)]$. Writing T for the S-set $\{z: \exists x.\, z = \tau(x)\}$, it follows that $\vdash_S$ *T is a chain in E*, and accordingly for some $c \in E^*$, $\vdash_S$ *c is the supremum of T*. We claim that c is the supremum of C in E^*. First, c is obviously an upper bound for C. And it is the least upper bound since, if $e \in E^*$ satisfies $\vdash_S c_i \leq e$ for

[1] This was first observed, in the context of Heyting-algebra-valued models of set theory, by Grayson [1975].

all $i \in I$, then $\vdash_S \tau(i^\wedge) \leq e$ for all $i \in I$, so that $\vdash_S \forall x.\ \tau(x) \leq e$ by the Generalization Principle. Therefore $\vdash_S c \leq e$ and so $c \leq^* e$. Thus E^* E is strongly inductive and so by **MZL** has a maximal element m.

We finally show that m is maximal in E, that is,

(1) $$\vdash_S \forall x{\in}E\ [m \leq x \rightarrow m = x].$$

Since S is near-standard, to establish this it suffices to show that, for any E-singleton U we have

(2) $$\vdash_S \forall x{\in}U\ [m \leq x \Rightarrow m = x].$$

Defining V to be the S-set $\{x{\in}U\colon m \leq x\}$, it is easily seen that (1) is equivalent to

(3) $$\vdash_S V \subseteq \{m\}.$$

Now consider $V' = V \cup \{m\}$. This is (S-derivably) a chain in E (recall that V is a singleton), and so has a supremum v. Clearly $\vdash_S m \leq v$, so the maximality of m in E^* gives $\vdash_S m = v$. It follows that

(4) $$\vdash_S x \in V \Rightarrow x \leq v \Rightarrow x \leq m.$$

But since $\vdash_S x \in V \rightarrow m \leq x$, (4) yields

$$\vdash_S x \in V \Rightarrow x = m,$$

i.e. (2). ■

If H is a complete Heyting algebra, then, as we shall see below, $Th(\mathcal{S}et_H)$ satisfies the conditions placed on S in Prop. 8, so that, if **MZL** holds in the underlying set theory, it holds in $Th(\mathcal{S}et_H)$. Since the algebra of truth values of $Th(\mathcal{S}et_H)$ is isomorphic to H, it follows that **MZL** *is compatible with any intuitionistic algebra of truth values.* This must also be the case for

ZL since, as shown in Chapter II, **MZL** and **ZL** are equivalent in intuitionistic set theory. In particular, **ZL** *can have no nonconstructive logical consequences*[1]. This is in sharp contrast with **AC**, which, as we have seen, implies **LEM**.[2]

Since **MZL** has no nonconstructive logical consequences, **AC** in its usual form cannot be an intuitionistic consequence of it. But there is a weaker version of **AC** which can be shown to follow intuitionistically from **MZL.** This weaker version may be stated as follows. Given S-sets F, X, Y, let us say that F is a *partial function on X to Y* if

$$\vdash_S F \subseteq X \times Y \wedge \forall x \in X\, \forall y,z \in Y\, (\langle x, y\rangle \in F \wedge \langle x, z\rangle \in F) \Rightarrow y = z.$$

Then S is *weakly choice* if the following rule is valid:

$$\frac{\vdash_S \forall x \in X\, \exists y \in Y\, \alpha(x, y)}{\vdash_S \forall x{\in}X \forall y{\in}Y[\langle x,y\rangle \in M \Rightarrow \alpha(x,y)] \wedge X - \mathit{domain}\,(M)^3 = \varnothing}$$

for some partial function M from X to Y

An S-set M satisfying this condition is a partial choice function for α which is "almost" a full choice function in that the double complement of its domain coincides with X.

Now we can prove

Proposition 9. *If S is near-standard and Zornian, then S is weakly choice.*

Proof. Assume the hypotheses and

---

[1] For this reason it is very much more difficult to establish the independence of **ZL** from pure local set theory than that of **ZL**. Indeed, the only way seems to be to invoke the fact—which, as we have seen in Chapter IV, is comparatively difficult to prove — that its classical equivalent **AC** is independent of classical set theory.

[2] As mentioned in Chapter II, **ZL** was originally introduced in order to avoid the "transcendental" devices associated with the use of **AC.** That **ZL** is compatible with constructive reasoning provides unexpected further confirmation of its "non-transcendental" character.

[3] Here domain(M) is the S-set $\{x : \exists y(\langle x, y\rangle \in M\}$.

(1) $\vdash_S \forall x \in X\, \exists y \in Y\, \alpha(x, y)$.

Let $R = \{<x, y> \in X \times Y: \alpha(x, y)\}$ and $E = \{U: U \subseteq R \wedge \text{Fun}(U)\}$. Then $\vdash_S$ *$(E, \subseteq)$ is a partially ordered set* and the usual argument involving unions of chains can be applied to $(E, \subseteq)$ in S, to yield $\vdash_S$ *$(E, \subseteq)$ is strongly inductive.* Since S is Zornian, E has a maximal element M. Clearly $\vdash_S Fun(M) \wedge \forall x \in X \forall y \in Y[<x, y> \in M \Rightarrow \alpha(x,y)]$. To complete the proof we need to show that

(2) $\vdash_S X - domain(M) = \varnothing$.

To do this we argue informally in S. Suppose $a \in X$ - domain(M). Then from (1) it follows that there is $b \in Y$ for which $\alpha(a, b)$. Then $M' = M \cup \{<a, b>\}$ is a member of E containing M. Since M is maximal, $M = M'$, whence $<a, b> \in M$. This contradicts the assumption $a \in X$ - domain(M). It follows that X - domain(M) must be empty, i.e. (2). ■

Corollary. *Assuming* **AC,** *any full well-termed classical near-standard local set theory is choice.*

Proof. Let S be a full well-termed classical near-standard local set theory. Given **AC,** we then have **MZL,** so it follows from Prop. 8 that S is Zornian. From Prop. 9 we deduce that S is weakly choice. But clearly any classical weakly choice local set theory is choice, and the conclusion follows. ■

THE FOREGOING PRINCIPLES INTERPRETED IN TOPOSES

When S is the theory $Th(\mathscr{E})$ of a topos $\mathscr{E}$, the conditions on S formulated in the previous section are correlated with certain properties of $\mathscr{E}$, which we now proceed to determine.

$\mathscr{E}$ is said to be *extensional* provided that, for any objects A, B and any pair of arrows $A \xrightarrow{f} B, A \xrightarrow{g} B$, if $fh = gh$ for every arrow $1 \xrightarrow{h} A$, then $f = g$. We recall that this says that each object of $\mathscr{E}$ satisfies the Axiom of Extensionality in the sense that its identity as a domain is entirely determined by its "elements".

A weaker version of extensionality is obtained by replacing 1 with subobjects of 1, that is, objects U for which the unique arrow $U \to 1$ is monic. Thus $\mathscr{E}$ is said to be *subextensional* provided that for any objects A, B and any pair of arrows $A \xrightarrow{f} B, A \xrightarrow{g} B$, if $fh = gh$ for every $U \xrightarrow{h} A$ with $U \rightarrowtail 1$, then $f = g$.

We recall that a category is said to satisfy the Axiom of Choice (**AC**) if , for any epic $f: A \twoheadrightarrow B$, there is a (necessarily monic) $g: B \to A$ such that $fg = 1_B$, or equivalently, if each of its objects is projective. It satisfies the Strong Axiom of Choice (**SAC**) if for any object $X \not\cong 0$ and any arrow $f: X \to Y$, there is an arrow $g: Y \to X$ such that $fgf = f$.

$\mathscr{E}$ is *Boolean* if the arrow $1+1 \xrightarrow{\top + \bot} \Omega$ is an isomorphism, and *bivalent* if $\top$ and $\bot$ are the only arrows$1 \to \Omega$, or equivalently, 1 has only the two subobjects 0 and 1.

Let A be an object of $\mathscr{E}$, and let $m: B \rightarrowtail A$ be a subobject of A. A *complement* for B is a subobject $n: C \rightarrowtail A$ such that the arrow $m + n: B + C \to A$ is an isomorphism. Then it is easy to show that *$\mathscr{E}$ is Boolean if and only if every object in $\mathscr{E}$ has a complement.*

Notice that, even if we only assume intuitionistic logic in our metatheory, $\mathcal{S}et$ is extensional. If full classical logic is assumed, $\mathcal{S}et$ is both Boolean and bivalent.

If S is a well-endowed local set theory, and $\mathscr{E}$ is a topos, we have the following concordance between properties of S (respectively $Th(\mathscr{E})$) and properties of $\mathscr{C}(S)$ (respectively $\mathscr{E}$):

S, $Th(\mathscr{E})$	$\mathscr{C}(S)$, $\mathscr{E}$
CONSISTENT	NONDEGENERATE
CLASSICAL	BOOLEAN
COMPLETE	BIVALENT
STANDARD	EXTENSIONAL
NEAR-STANDARD	SUBEXTENSIONAL
WITNESSED	1 IS PROJECTIVE
CHOICE	SATISFIES **AC**
HILBERTIAN	SATISFIES **SAC**
FULL	WELL-COPOWERED

We prove a couple of these equivalences, leaving the rest to the reader.

If S is well-endowed, then S is standard iff $\mathscr{C}(S)$ is extensional. If S is well-endowed, then $\mathscr{C}(S)$ is equivalent to the category $\mathscr{T}(S)$ of S-types and terms, so to establish the extensionality of $\mathscr{C}(S)$ it is enough to establish that of $\mathscr{T}(S)$. Accordingly let **A, B** be type symbols and suppose that $f, g: A \to B$ are $\mathscr{T}(S)$-arrows such that, for any $\mathscr{T}(S)$-arrow $1 \xrightarrow{h} A$, we have $fh = gh$. Now f is $x \mapsto \xi$ and g is $x \mapsto \eta$ for some terms ξ, η, and the condition just stated becomes: for any closed term τ of type **A**, we have $\vdash_S \xi(\tau) = \eta(\tau)$.

Supposing that S is standard, it follows that $\vdash_S \forall x(\xi(x) = \eta(x))$, whence $f = g$. So $\mathcal{T}(S)$, and hence also $\mathcal{C}(S)$, is extensional.

Conversely, suppose $\mathcal{C}(S)$ is extensional. Let **A** be a type symbol and $\alpha(x)$ a formula with a free variable of type **A**. Let f be the S-map $(x \mapsto \alpha)$: $A \to \Omega$. If $\vdash_S \alpha(\tau)$ for all closed terms τ of type **A**, it follows that the diagram

$$1 \xrightarrow{\ \tau\ } A \underset{T_A}{\overset{f}{\rightrightarrows}} \Omega$$

commutes for all such τ. Since $\mathcal{C}(S)$ is extensional (and well-termed), we deduce that $f = T_A$, in other words that $\vdash_S \forall x(\alpha(x) = \top)$, i.e. $\vdash_S \forall x\alpha(x)$. So S is standard.

S is choice iff $\mathcal{C}(S)$ satisfies **AC.** Given an epic g: $Y \twoheadrightarrow X$ in $\mathcal{C}(S)$, let α be the formula $\langle y, x\rangle \in g$. Then $\vdash_S \forall x{\in}X\exists y{\in}Y\alpha(x,y)$. If S is choice there is f: $X \to Y$ such that $\vdash_S \forall x{\in}X\alpha(x,fx)$, from which it follows easily that $gf = 1_X$. So $\mathcal{C}(S)$ satisfies **AC**.

Conversely, suppose $\mathcal{C}(S)$ satisfies **AC** and $\vdash_S\forall x{\in}X\exists y{\in}Y\alpha(x,y)$ for a given formula α. Define $Z = \{\langle x,y\rangle \in X \times Y: \alpha\}$ and $g = (\langle x,y\rangle \mapsto x)$: $Z \to X$, $k = (\langle x,y\rangle \mapsto y)$: $Z \to Y$. Then g is epic, and so by **AC** there is h: $X \to Z$ such that $gh = 1_X$. If we now define $f = kh$: $X \to Y$, it is easy to see that $\vdash_S \forall x{\in}X\alpha(x,fx)$. So S is choice.

It follows from this that *any topos satisfying* **AC** *is Boolean*, so that subobjects always possess complements.

Remark. The original proof that any topos satisfying **AC** is Boolean is based on the idea of constructing a complement for any subobject. Here is a highly informal version of the argument.

Suppose that the topos satisfies **AC**, and let X be a subobject of an object A. Form the coproduct $A + A$, and think of it as the union of two disjoint copies of A. Regard the elements of the first copy as being coloured black and those of the second as being coloured white. Thus each element of A has been 'split' into a 'black' copy and a 'white' copy. Next, identify each copy of an element of X in the first (black) copy with its mate in the second (white) copy; the elements thus arising we agree to colour grey, say. In this way we obtain a set Y consisting of black, white and grey elements[1], together with an epic map $A + A \twoheadrightarrow Y$. Now we use **AC** to assign each element $y \in Y$ an element $y' \in A + A$ in such a way that y' is sent to y by the map $A + A \twoheadrightarrow Y$ above. The whole process—call it P, say—accordingly transforms each element of $A + A$ into an element (possibly the same) of $A + A$. Now, for $n = 0, 1, 2$, define

$A_n = \{a \in A$: *P effects a change in colour in exactly n copies of a*$\}$.

Then clearly $A = A_0 \cup A_1 \cup A_2$, $A_1 = X$ and $A_2 = \varnothing$. It follows that A_0 is a complement for X.

Some examples[2].

(i) *$\mathcal{S}et$ is extensional, satisfies* **AC§**, *and is both Boolean§ and bivalent§*

(ii) *For any partially ordered set P, $\mathcal{S}et^P$ is subextensional . It satisfies AC iff§, and only if, P is trivially ordered, that is, if the partial ordering*

---

[1] One should not be misled into thinking that *at this stage* the 'grey' elements of Y can be clearly distinguished from the 'black' and 'white' ones: since the former are correlated with the elements of X, such distinguishability would be tantamount to assuming that X already possesses a complement!

[2] In presenting these examples we indicate by appending the symbol **§** when we need to assume that Set satisfies **AC**, or at least that its internal logic is classical and bivalent.

in P coincides with the identity relation. To show that $\mathcal{S}et^P$ is subextensional, given $\alpha, \beta: F \to G$ in $\mathcal{S}et^P$, $p_0 \in P$ and $a \in F(p_0)$, define $U \in \mathcal{S}et^P$ by $U(p) = \{x: x = 0 \wedge p_0 \leq p\}$ with the U_{pq} the obvious maps. Then U is a subobject of 1 in $\mathcal{S}et^P$. Define $\varphi: U \to F$ by $\varphi_p = U(p) \times \{a\}$. If $\alpha\varphi = \beta\varphi$, then $\alpha_{p_0} \circ \varphi_{p_0}(0) = \beta_{p_0} \circ \varphi_{p_0}(0)$, whence $\alpha_{p_0}(a) = \beta_{p_0}(a)$. Since p_0 and a were arbitrary, $\alpha = \beta$. So $\mathcal{S}et^P$ is subextensional.

To show that **AC** holds in $\mathcal{S}et^P$ only if **P** is trivially ordered, suppose that $p_0 < q_0$ in **P** and define A, B in $\mathcal{S}et^P$ by $A(p) = \{0, 1\}$ for all $p \in P$, and each A_{pq} the identity map; $B(p) = \{0\}$ if $p_0 < p$, $B(p) = \{0,1\}$ if $p_0 \not< p$, each B_{pq} either the identity map on $\{0,1\}$ or the map $\{0,1\} \to \{0\}$ as appropriate. Then it is easy to show that the map $f: A \to B$ in $\mathcal{S}et^P$ —with each f_p either the identity map on $\{0,1\}$ or the map $\{0,1\} \to \{0\}$ as appropriate—has no section.

(iii) *For any complete Heyting algebra H, $\mathcal{S}et_H$ is subextensional. It satisfies **AC** if§, and only if, H is a Boolean algebra*[1]. To show that $\mathcal{S}et_H$ is subextensional, suppose given $f, g: (I, \delta) \to (J, \varepsilon)$ in $\mathcal{S}et_H$. For $i_0 \in I$, $j_0 \in J$, let $\eta_i = g_{i_0 j_0} \wedge \delta_{i i_0}$ and $a = \bigvee_{i \in I} \eta_i$. Then $(\{0\}, \lambda)$ with $\lambda_{00} = a$ is a subobject of 1 in $\mathcal{S}et_H$ and the η_i define an arrow $\eta: (\{0\}, \lambda) \to (I, \delta)$. If $f\eta = g\eta$, then a calculation shows that $f_{i_0 j_0} = g_{i_0 j_0}$. Since i_0 and j_0 were arbitrary, $f = g$.

As for the second contention, if $\mathcal{S}et_H$ satisfies **AC**, it is Boolean, and so H must be a Boolean algebra. Conversely, if H is a Boolean algebra, then $\mathcal{S}et_H$ is Boolean, so $Th(\mathcal{S}et_H)$ is classical. It is

[1] If B is a complete Boolean algebra, $\mathcal{F}uz_B$ is equivalent to $\mathcal{S}et_B$, so **AC** also holds in $\mathcal{F}uz_B$.

not hard to show that $\mathcal{S}et_H$ has all set-indexed copowers of 1, so that $Th(\mathcal{S}et_H)$ is full. We also know that $\mathcal{S}et_H$ is subextensional, so that $Th(\mathcal{S}et_H)$ is near-standard. It follows from the Corollary to Prop. 8. that $Th(\mathcal{S}et_H)$ is choice, so that $\mathcal{S}et_H$ satisfies **AC.**

(iv)§ *For a monoid* **M**, *the topos* $\mathcal{S}et^{\mathbf{M}}$ *of* **M**-*sets is bivalent.* For the terminal object in $\mathcal{S}et^{\mathbf{M}}$ is the one-point set 1 with trivial **M**-action and evidently this has only the two subobjects 0, 1.

(v) For *a monoid* **M**, *if the topos* $\mathcal{S}et^{\mathbf{M}}$ *is Boolean, then* **M** *is a group*[1], *and conversely*§. For suppose that $\mathcal{S}et^{\mathbf{M}}$ is Boolean. Regard **M** as an **M**-set with the natural multiplication on the left by elements of **M.** For $a \in M$, $U = \{xa: x \in M\}$ is a sub-**M**-set of **M**, and so has a complement V in $\mathcal{S}et^{\mathbf{M}}$ which must itself be an sub-**M**-set of **M.** Now $1 \notin V$, since otherwise $V = M$ which would make U empty. It follows that $1 \in U$ and so a has a left inverse. Since any monoid with left inverses is a group, **M** is a group. Conversely, if **M** is a group (and $\mathcal{S}et$ is Boolean), then the set-theoretical complement of any sub-**M**-set Y of an **M**-set X is itself a sub-**M**-set and therefore the complement in $\mathcal{S}et^{\mathbf{M}}$ of Y.

(vi) *If* **G** *is a nontrivial group, then 1 is not projective in* $\mathcal{S}et^{\mathbf{G}}$. For $G \to 1$ in $\mathcal{S}et^{\mathbf{G}}$ is epic, but an arrow $1 \to G$ in $\mathcal{S}et^{\mathbf{G}}$ corresponds to an element $e \in G$ such that $ge = e$ for all $g \in G$, which cannot exist unless G has just one element.

(vii) *For a monoid* **M**, $\mathcal{S}et^{\mathbf{M}}$ *satisfies* **AC** *if*§, *and only if*, **M** *is trivial.* If $\mathcal{S}et^{\mathbf{M}}$ *satisfies* **AC,** then $\mathcal{S}et^{\mathbf{M}}$ is Boolean and so by **(v)** **M** is a group. But by **(vi)** if **M** is nontrivial, 1 is not projective in $\mathcal{S}et^{\mathbf{M}}$, and so $\mathcal{S}et^{\mathbf{M}}$ does not satisfy **AC.** It follows that **M** is trivial.

[1] It follows that if **M** is not a group, then $\mathcal{S}et^{\mathbf{M}}$ is bivalent§ but not Boolean.

CHARACTERIZATION OF $\mathcal{S}et$

We remind the reader that we are assuming that our background metatheory is constructive. For definiteness we will take that metatheory to be intuitionistic Zermelo-Fraenkel set theory **IZF.** Now consider the topos $\mathcal{S}et$ in **IZF**. We seek to determine necessary and sufficient conditions on a local set theory S for its associated topos of sets $\mathscr{C}(S)$ to be equivalent, as a category, to $\mathcal{S}et$. We shall see that the conjunction of standardness with a new property, fullness, meets the requirements. Moreover, if we replace **IZF** by classical ZF and in addition assume that $\mathcal{S}et$ satisfies **AC**, then the conjunction of fullness, choiceness, and completeness, as well as the conjunction of and fullness and Hilbertianness also works.

We can now prove the

Theorem.

Let S be a full well-endowed consistent local set theory S. Then

(i) *the following are equivalent:*

(*a*) $\mathscr{C}(S) \simeq \mathcal{S}et$.

(*b*) *S is standard,*

(ii) *Assuming both classical logic in the metatheory and that* $\mathcal{S}et$ *satisfies* **AC**, *conditions (a) and (b) are each equivalent to*

(*c*) *S is choice and complete,*

(*d*) *S is Hilbertian.*

Proof. (i) Assuming (*a*), we note that since $\mathcal{S}et$ is extensional and has arbitrary set-indexed copowers of 1, so does $\mathscr{C}(S)$. But then S is standard and full, i.e. (*b*).

For the converse, suppose that S is full. Since S is well-termed, for any S-map $f\colon X \to Y$ we can write $f(\tau)$ for each closed term τ such that $\vdash_S \tau \in X$.

We define functors Δ: $\mathscr{C}(S) \to \mathcal{S}et$, ^: $\mathcal{S}et \to \mathscr{C}(S)$, which, under the specified conditions, we show defines an equivalence.

First, $\Delta(X)$ is the set of closed terms τ such that $\vdash_S \tau \in X$, where we identify σ, τ if $\vdash_S \sigma = \tau$. Given f: $X \to Y$, we define $\Delta(f)$ to be the map $(\tau \mapsto f(\tau))$: $\Delta(X) \to \Delta(Y)$.

Next, given I in $\mathcal{S}et$, we define I^ to be the S-set $U_{I\wedge}$. Given f: $I \to J$, there is a term f^ : **J^** ith x **: I^** such that $\vdash_S f\text{^}(i\text{^}) = (fi)\text{^}$ for all $i \in I$. We define f^: I^ $\to$ J^ to be the S-map $x \mapsto f\text{^}(x)$. It is easily shown, using the Generalization Principle, that, for $I \xrightarrow{f} J \xrightarrow{g} K$, $\vdash_S (g \circ f)\text{^} = g\text{^} \circ f\text{^}$. Moreover, if f^ is epic in $\mathscr{C}(S)$, then f is epic in $\mathcal{S}et$. For suppose $J \xrightarrow{g} K$ and $J \xrightarrow{h} K$ satisfy $g \circ f = h \circ f$. Then

$$\vdash_S g\text{^} \circ f\text{^} = (g \circ f)\text{^} = (h \circ f)\text{^} = h\text{^} \circ f\text{^},$$

so if f^ is epic in C(S), it follows that $\vdash_S g\text{^} = h\text{^}$.. Hence, for each $i \in I$, $\vdash_S (gi)\text{^} = g\text{^}(i\text{^}) = h\text{^}(i\text{^}) = (hi)\text{^}$, so that $gi = hi$ for each $i \in I$, that is, $g = h$. Thus f is epic.

For any set I and any S-set X, we have natural maps η: $I \to \Delta(I\text{^})$ and ε: $\Delta(X)\text{^} \to X$ defined as follows:

$$\eta_I(i) = i\text{^} \text{ for } i \in I; \quad \vdash_S \varepsilon(\tau\text{^}) = \tau \text{ for all } \tau \in \Delta(X).$$

Clearly η is monic. The same is true of ε since for $\sigma, \tau \in \Delta(X)$,

$$\vdash_S \varepsilon(\sigma\text{^}) = \varepsilon(\tau\text{^}) \Rightarrow \sigma = \tau,$$

whence

$$\vdash_S \forall x \forall y[\varepsilon(x) = \varepsilon(y) \Rightarrow x = y]$$

by the Generalization Principle.

Now suppose that S is also standard. We claim that then ε is epic and hence an isomorphism. For we have, for all $\tau \in \Delta(X)$,

$\vdash_S \varepsilon(\hat{\tau}) = \tau$, whence $\vdash_S \exists y\varepsilon(y) = \tau$. Since X is standard, we infer that

$$\vdash_S \forall x{\in}X\exists y\ \varepsilon(y) = x,$$

so that ε is onto, hence epic.

Using the fact that ε is an isomorphism we can now show that η is epic, and hence also an isomorphism. For consider $\eta\wedge : I\wedge \to (\Delta I\wedge)\wedge$. We note that

(*) $$\varepsilon \circ \eta\wedge = 1_{I\wedge}.$$

For if $i \in I$, then

$$\vdash_S \varepsilon(\eta \wedge (i\wedge)) = \varepsilon((\eta i)\wedge) = \eta i = i\wedge.$$

It follows by the Generalization Principle that

$$\vdash_S \forall x \in I\wedge.\ \varepsilon(\eta x) = x,$$

whence (*).

Since ε is an isomorphism, it follows easily from (*) that $\eta\wedge$ is an isomorphism, hence also epic. Accordingly η is itself epic, and hence also an isomorphism.

We conclude that (Δ, ^) define an equivalence between $\mathscr{C}(S)$ and $\mathcal{S}et$, as required.

(ii) We have already shown (Prop. 6) that (*c*) ⇒ (*b*). Now let S satisfy (*a*), that is, $\mathscr{C}(S) \simeq \mathcal{S}et$. Assuming classical logic in the metatheory, $\mathcal{S}et$, whence also $\mathscr{C}(S)$, is bivalent, so S is complete. Assuming that $\mathcal{S}et$ satisfies **AC**, $\mathscr{C}(S)$ does likewise, and so S is choice. Finally, $\mathcal{S}et$ has arbitrary set-indexed copowers of 1, so also then does $\mathscr{C}(S)$, and thus S is full. In other words we have shown (*a*) ⇒ (*c*). Finally (*c*) ⇔ (*d*) has been established in Prop. 7.

■

Using the concordance between properties of toposes and properties of local set theories, the previous theorem immediately yields the

Corollary. *Let $\mathscr{E}$ be a well-copowered topos. Then*

(i) *the following are equivalent:*

(a) $\mathscr{E} \simeq \mathcal{S}et$.

(b) $\mathscr{E}$ *is extensional,*

(ii) *Assuming both classical logic in the metatheory and that $\mathcal{S}et$ satisfies* **AC***, conditions (a) and (b) are each equivalent to*

(c) $\mathscr{E}$ is bivalent and satisfies **AC**

(d) $\mathscr{E}$ satisfies **SAC**.

Thus we see that $\mathcal{S}et$ is characterized up to equivalence by the fact that it is well-copowered and satisfies **SAC.**

It is also possible to formulate similar characterizations of other toposes, for example categories of presheaves over partially ordered sets, sheaves over topological spaces, and *H*-sets. For instance, a topos $\mathscr{E}$ is equivalent to $\mathcal{S}et_H$ for some complete Heyting algebra *H* if and only if $\mathscr{E}$ is well-copowered and subextensional and near-standard, and $\mathscr{E}$ is equivalent to $\mathcal{S}et_B$ for some complete Boolean algebra *B* if and only if $\mathscr{E}$ is well-copowered and satisfies **AC.**

VII
The Axiom of Choice in Constructive Type Theory

CONSTRUCTIVE TYPE THEORY

The roots of type theory lie in set theory, to be precise, in Bertrand Russell's efforts to resolve the paradoxes besetting set theory at the end of the 19th century. In the course of analyzing these paradoxes Russell had come to find the set, or class, concept itself philosophically perplexing, and the theory of types can be seen as the outcome of his struggle to resolve these perplexities. In Russell's initial conception of types, which later became known as the "simple" theory of types, the universe of logical objects is stratified into "layers" or "types", and each logical object is assigned a definite type. Relationships among objects must respect the types assigned to each object: thus, for example, two objects can be equal only if they have the same type, and one object can be a member of another object only if the type of the first object is the immediate predecessor of that of the second. Later, Russell came to regard the simple theory of types as inadequate for dealing with the more subtle "paradoxes of definition" which had appeared and so replaced it with the considerably more complicated system of "ramified" type theory which he and A. N. Whitehead developed in their *Principia Mathematica* of 1910-13. This monumental work embodies Russell's central logicist goal of reducing mathematics to logic.

In the form of the "Multiplicative Axiom" (essentially what we have called **CAC**) **AC** played a significant role in *Principia Mathematica.* Along with the Axiom of Infinity and the infamous

Axiom of Reducibility, **AC** was a member of the trio of "awkward" principles that Russell saw as necessary for the development of mathematics, but could not be justified on purely logical grounds[1]. Russell himself described **AC** as capable of being "enunciated, but not proved, in terms of logic."[2] In the 1920s Ramsey championed simple type theory, within which, he claimed, **AC** could be seen as "the most evident tautology". [3] In the early 1940s Church gave the definitive formulation of simple type theory in terms of the λ-calculus which is still standard today. But with respect to both of the original forms of type theory —ramified and simple—the status of **AC** was fundamentally no different from that it held with respect to set theory—namely, as a natural, even self-evident principle, but still, like the parallel postulate, undemonstrable.

Type theory took a remarkable turn in the 1980s with the emergence of the so-called *propositions-as-types doctrine* (or interpretation). Underlying this doctrine is the idealist notion, traceable to Kant, and central to Brouwerian intuitionism, that the meaning of a proposition does not derive from an absolute standard of truth external to the mind, but resides rather in the evidence for its assertability in the form of a mental construction or proof. Thus the central thesis of the propositions-as-types doctrine is that each proposition is to be *identified* with the type, set, or assemblage of its proofs[4]. As a result, such proof types, or sets of proofs, have to be accounted the *only* types, or sets. Strikingly, then, the propositions-as-types doctrine decrees that a

---

[1] The Axioms of Infinity, Reducibility and Choice were needed to develop arithmetic, real analysis, and set theory, respectively.

[2] Russell [1919]

[3] Ramsey [1926]

[4] This idea was advanced by Curry and Feys [1958] and later by Howard [1980]. As the *Curry-Howard correspondence* it has come to play an important role in theoretical computer science.

type, or set, simply *is* the type, or set, of proofs of a proposition, and, reciprocally, a proposition *is* just the type, or set, of its proofs. These are truly radical identifications. And remarkably, as we shall see, these identifications *render* **AC** *demonstrable.*

In the original type theories of Russell and Church, each type is independent of other types and is thus, so to speak, absolute or static; this holds in particular of the type of propositions or truth values. Now formulas or propositional functions in general manifest variation, since their values vary over, or depend on, the domain(s) of their free variables. Because of this they cannot be accurately represented as static types. This limitation makes it impossible for the earlier type theories to realize faithfully the propositions-as-types doctrine. In order to achieve this it is necessary to develop a theory of "variable" or *dependent* types, wherein types can depend on, or "vary over" other types. In a dependent type theory, type symbols may take the form $\mathbf{B}(x)$, with x a variable of a given type $\mathbf{A}$: $\mathbf{B}(x)$ is then a type dependent on or varying over the type **A.** The introduction of dependent types is also essential for the proper formulation of **AC** in conformity with the propositions-as-types doctrine.

Such a theory—*Constructive* (*Dependent*) *Type Theory* — was introduced[1] by Martin-Löf[2]. His theory, which has subsequently undergone much development, is also (as its name indicates) the first strictly constructive theory of types, in the sense of being both predicative (so in particular it lacks a type of propositions) and based on intuitionistic logic. In introducing it Martin-Löf's

[1] Dependent types were actually first studied in the late 1960s by de Bruijn and his colleagues at the University of Eindhoven in connection with the AUTOMATH project. Constructive type theory has been employed as a basis for various computational devices employed for the verification of mathematical theories and of software and hardware systems in computer science.

[2] Martin-Löf [1975], [1982], [1984].

purpose was to provide, as he put it[1] "a full scale system for formalizing intuitionistic mathematics as developed, for example, in the book by Bishop[2]." Martin-Löf's system provides a complete embodiment of the propositions-as-types doctrine[3]. Here is Martin-Löf himself on the latter[4]:

> *Every mathematical object is of a certain kind or type. Better, a mathematical object is always given together with its type, that is it is not just an object: it is an object of a certain type. ... A type is defined by prescribing what we have to do in order to construct an object of that type... Put differently, a type is well-defined if we understand...what it means to be an object of that type. ... Note that it is required, neither that we should be able to generate somehow all the objects of a given type, nor that we should, so to say, know all of them individually. It is only a question of understanding what it means to be an arbitrary object of the type in question.*
>
> *A proposition is defined by prescribing how we are allowed to prove it, and a proposition holds or is true intuitionistically if there is a proof of it. ... Conversely, each type determines a proposition, namely, the proposition that the type in question is nonempty. This is the proposition which we prove by exhibiting an object of the type in question. On this analysis, there appears to be no fundamental difference between propositions and types. Rather, the difference is one of point of view: in the case of a proposition, we are not so much interested in what its proofs are as in whether it has a proof, that is, whether it is true or false, whereas, in the case of a type, we are of course interested in what its objects are and not only in whether it is empty or nonempty.*

---

[1] Martin-Löf [1975].

[2] I.e. Bishop [1967].

[3] Martin-Löf's original calculus contained a type of all types. This assumption was shown to be inconsistent by Girard [1972]. Martin-Löf accordingly dropped this assumption in later versions of his theory.

[4] Martin-Löf [1975].

The propositions-as-types doctrine gives rise to a correspondence between logical operators and operations on (dependent) types. Let us follow Tait's exposition[1] of the idea in set-theoretic terms. To begin with, consider two propositions/types/sets **A** and **B**. What should be required of a proof f of the implication $\mathbf{A} \to \mathbf{B}$? Simply that, given any proof x of **A**, f should yield a proof of **B**, that is, f should be a function from **A** to **B**. In other words, the proposition $A \to B$ is just the type of functions from **A** to **B**:

$$\mathbf{A} \to \mathbf{B} = \mathbf{B}^{\mathbf{A}}$$

Similarly, all that should be required of a proof c of the conjunction $\mathbf{A} \wedge \mathbf{B}$ is that it should yield proofs x and y of **A** and **B**, respectively. From this point of view $\mathbf{A} \wedge \mathbf{B}$ is accordingly just the type $\mathbf{A} \times \mathbf{B}$ – the *product* **A** and **B**—of pairs (x, y), with x of type **A** (we write this as x: **A**) and y: **B**.

A proof of the disjunction $\mathbf{A} \vee \mathbf{B}$ is either a proof of **A** or a proof of **B** together with the information as to which of **A** or **B** it is a proof. That is, if we introduce the type **2** with the two distinct elements 0 and 1, a proof of $\mathbf{A} \vee \mathbf{B}$ may be identified as a pair (c, n) in which either c is a proof of **A** and n is 0, or c is a proof of **B** and n is 1. This means that $\mathbf{A} \vee \mathbf{B}$ should be construed as the type of such pairs, that is, the *two-term dependent sum* $\mathbf{A} + \mathbf{B}$ of **A** and **B**.

The true proposition $\top$ may be identified with the one element type $\mathbf{1} = \{0\}$: 0 thus counts as the unique proof of $\top$. The false proposition $\bot$ is taken to be a proposition which lacks a proof altogether: accordingly $\bot$ is identified with the empty set $\varnothing$. The negation $\neg\mathbf{A}$ of a proposition **A** is defined as $\mathbf{A} \to \bot$, which therefore becomes identified with the set $\mathbf{A}^{\varnothing}$.

[1] Tait [1994].

As we have already said, a proposition A is deemed to be true if it (i.e, the associated type **A**) has an element, that is, if there is a function $\mathbf{1} \to \mathbf{A}$. Accordingly the *Law of Excluded Middle* for a proposition A becomes the assertion that there is a function $\mathbf{1} \to A + \varnothing^A$.

If a and b are objects of type **A**, we introduce the *identity proposition* or *type* $a =_{\mathbf{A}} b$ expressing that a and b are identical objects of type **A**. This proposition is true, that is, the associated type has an element, if and only if a and b are identical. Here the term "identical" is to be taken in the *intensional* sense of affirming a literal identity of the two objects in question, rather than the extensional meaning the term receives in set theory, where two sets are taken as identical if they have the same members.

In order to deal with the quantifiers we require operations defined on families of types, that is, types $\mathbf{\Phi}(x)$ depending on objects x of some type **A**. By analogy with the case $\mathbf{A} \to \mathbf{B}$, a proof f of the proposition $\forall x{:}\mathbf{A}\ \mathbf{\Phi}(x)$, that is, an object of type $\forall x{:}\mathbf{A}\ \mathbf{\Phi}(x)$, should associate with each $x{:}\ \mathbf{A}$ a proof of $\mathbf{\Phi}(x)$. So f is just a function with domain A such that, for each $x{:}\ \mathbf{A}$, fx is of type $\mathbf{\Phi}(x)$. Accordingly, $\forall x{:}\mathbf{A}\ \mathbf{\Phi}(x)$ is the type of such functions, that is, the *dependent product* $\Pi x{:}\mathbf{A}\ \mathbf{\Phi}(x)$ of the $\mathbf{\Phi}(x)$'s. We use the λ-notation in writing f as $\lambda x fx$.

A proof of the proposition $\exists x{:}\mathbf{A}\ \mathbf{\Phi}(x)$, that is, an object of type $\exists x{:}\mathbf{A}\ \mathbf{\Phi}(x)$, should determine an object $x{:}\ \mathbf{A}$ and a proof y of $\mathbf{\Phi}(x)$, and *vice-versa.* So a proof of this proposition is just a pair (x, y) with $x{:}\ \mathbf{A}$ and $y{:}\ \mathbf{\Phi}(x)$. Therefore $\exists x{:}\mathbf{A}\ \mathbf{\Phi}(x)$ is the type of such pairs, that is, the *dependent sum* $\coprod x{:}\mathbf{A}\ \mathbf{\Phi}(x)$ of the $\mathbf{\Phi}(x)$'s.

To translate all this into the language of Constructive Type Theory[1], one uses the following concordance among operations:

Logical	Set-theoretic	Type-theoretic
$\wedge$	$\times$	$\mathbf{\times}$
$\vee$	$+$	two-term dependent sum
$\Rightarrow$	set exponentiation	type exponentiation
$\forall x$	cartesian product $\prod_{i\in I}$	dependent product $\Pi x{:}\mathbf{A}$
$\exists x$	disjoint union $\coprod_{i\in I}$ [2]	dependent sum $\coprod x{:}\mathbf{A}$

AC IN CONSTRUCTIVE TYPE THEORY

We now turn to the expression of **AC** in Constructive Type Theory. Again following Tait, we introduce the functions σ, π, π' of types $\forall x{:}\mathbf{A}(\mathbf{\Phi}(x) \to \exists x{:}\mathbf{A}\ \mathbf{\Phi}(x))$, $\exists x{:}\mathbf{A}\ \varphi(x) \to \mathbf{A}$, and $\forall y{:}\ (\exists x\ \mathbf{\Phi}(x)).\ \mathbf{\Phi}(\pi(y))$ as follows. If $b{:}\ \mathbf{A}$ and $c{:}\ \mathbf{\Phi}(b)$, then σbc is (b, c). If $d{:}\ \exists x{:}\mathbf{A}\ \mathbf{\Phi}(x)$, then d is of the form (b, c) and in that case $\pi(d) = b$ and $\pi'(d) = c$. These yield the equations

$$\pi(\sigma bc) = b \quad \pi'(\sigma bc) = c \quad \sigma(\pi d)(\pi' d) = d.$$

We shall use the following version of **AC3** to represent the Axiom of Choice—the *type-theoretic* Axiom of Choice:

ACT $\quad \forall x{:}\mathbf{A}\exists y{:}\mathbf{B}\ \mathbf{\Phi}(x, y)) \to \exists f{:}\mathbf{B}^{\mathbf{A}}\forall x{:}\mathbf{A}\ \mathbf{\Phi}(x, fx)).$

We shall now show that **ACT** is *provable* in constructive type theory, and accordingly *correct* under the propositions as types doctrine. For let u be a proof of the antecedent $\forall x{:}\mathbf{A}\exists y{:}\mathbf{B}\ \mathbf{\Phi}(x, y))$. Then, for any $x{:}\ \mathbf{A}$, $\pi(ux)$ is of type $\mathbf{B}$ and $\pi'(ux)$ is a proof

[1] For a complete specification of the operations and rules of constructive type theory, see Chapter 10 of Jacobs [1999] or Gambino and Aczel [2005].

[2] In set theory the disjoint union $\coprod_{i\in I} A_i$ of a family of sets $\{A_i{:}\ i \in I\}$ is defined to be the set $\bigcup_{i\in I} A_i \times \{i\}$.

of $\mathbf{\Phi}(x, \pi ux)$. So $s(u) = \lambda x.\pi(ux)$ is of type $\mathbf{B}^{\mathbf{A}}$ and $t(u) = \lambda x.\ \pi'(ux)$ is a proof of $\forall x{:}\mathbf{A}\ \mathbf{\Phi}(x, s(u)x)$. Accordingly $\lambda u.\sigma s(u)t(u)$ is a proof of $\forall x{:}\mathbf{A}\exists y{:}\mathbf{B}\ \mathbf{\Phi}(x, y)) \to \exists x{:}\mathbf{B}^{\mathbf{A}}\ \forall x{:}\mathbf{A}\ \mathbf{\Phi}(x, fx))$. This proves **ACT.**

Put informally, what this shows is that in Constructive Type Theory the consequent of **ACT** means *nothing more than its antecedent.* Indeed, as we have already pointed out, from a strictly constructive point of view, the assertability of an alternation of quantifiers $\forall x\exists yR(x,y)$ means *precisely* that one is given a function f for which $R(x, fx)$ holds for all x.

What does the above derivation of **ACT** amount to in set-theoretic terms? Tracing the argument through using the set-theoretic column of the above concordance, one finds that, rather than demonstrating **AC** in any of its set-theoretic forms, it establishes a bijection, for any doubly indexed family of sets $\{A_{ij}: \langle i,j\rangle \in I \times J\}$, between the sets $\prod_{i\in I}\coprod_{j\in J} A_{ij}$ and $\coprod_{f\in J^I}\prod_{i\in I} A_{ifi}$. This bijection is natural and does not require the use of **AC** to prove its existence set-theoretically. On the other hand, in set theory **AC** is not represented by this, or any other, bijection, but rather by each of the two *equalities* in which $\coprod$ is replaced by $\bigcup$, and Π is replaced by $\bigcap$ in one, but not the other. These are the distributive laws

$$\bigcap_{i\in I}\bigcup_{j\in J} A_{ij} = \bigcup_{f\in J^I}\bigcap_{i\in I} A_{ifi} \qquad \prod_{i\in I}\bigcup_{j\in J} A_{ij} = \bigcup_{f\in J^I}\prod_{i\in I} A_{ifi}.$$

These facts can be tabulated as follows:

Statement	Type-theoretic interpretation	Set-theoretic interpretation
$\forall i\exists j\ A(i, j)$	$\prod_{i\in I}\coprod_{j\in J} A_{ij}$	$\bigcap_{i\in I}\bigcup_{j\in J} A_{ij}$ or $\prod_{i\in I}\bigcup_{j\in J} A_{ij}$

$\exists f \forall i\, A(i, fi)$	$\coprod_{f \in J^I} \prod_{i \in I} A_{ifi}$	$\bigcup_{f \in J^I} \bigcap_{i \in I} A_{ifi}$ or $\bigcup_{f \in J^I} \prod_{i \in I} A_{ifi}$

The presence of the natural bijection between the type-theoretic interpretations of $\forall i \exists j\, A(i, j)$ and $\exists f \forall i\, A(i, j)$ embodies the idea that, from the constructive standpoint, the two statements are not just logically equivalent but *intensionally* equivalent in that they *have the same meaning*[1]: the assertability of $\forall i \exists j\, A(i, j)$ means *precisely* to be given a function f for which $A(i, fi)$ holds for all i. The bijection, as it were, *converts* each element of the set representing $\forall i \exists j\, A(i, j)$ into an element of the set representing $\exists f \forall i\, A(i, j)$ (and vice-versa). Nothing further is required, under the propositions-as-types doctrine, to affirm the equivalence of the two statements. This equivalence has accordingly come to be termed the *intensional* Axiom of Choice: it is essentially tautologous[2], mathematically "trivial". By contrast, the equivalence between $\forall i \exists j\, A(i, j)$ and $\exists f \forall i\, A(i, j)$ as asserted by **AC** is represented, under the set-theoretic[3] interpretation, by the *extensional* equality of the representing sets, i.e., the assertion that,

[1] Here we again recall Bishop's [1967] assertion that *a choice function exists in constructive mathematics because a choice is implied by the very meaning of existence.*

[2] Precisely as Ramsey (*v . supra*) asserted, but in this case for quite different reasons. Ramsey construed, and accepted the truth of **AC** as asserting the objective existence of choice functions, given extensionally and so independently of the manner in which they might be described. But the intensional nature of constructive mathematics, and, in particular, of the "propositions-as-types" doctrine decrees that nothing is given completely independently of its description. This leads to a strong construal of the quantifiers which, as we have observed, "trivializes" **AC** by rendering the antecedent of the implication constituting it essentially equivalent to the consequent. It is remarkable that **AC** has been considered tautological both from an extensional and from an intensional point of view.

[3] Or topos-theoretic: see Chapter VI.

as a matter of fact, they have the same elements. This, **AC** as understood by the majority of mathematicians, has come to be called by type-theorists the *extensional* Axiom of Choice. From the standpoint of the practicing mathematician the extensional Axiom of Choice is *nontrivial* in the sense that its affirmation is more than a mere matter of definition. That being the case, it might be appropriate to call it the *Postulate*, rather than the *Axiom* of Choice, in accordance with the Greek mathematicians' use of the term "axiom" to signify a self-evident assertion, a universal assumption, while the term "postulate" was used for an assertion lacking such universality and applying only to the subject under study[1].

It is of interest to compare all this with the analysis of **AC** presented by Paul Bernays in the 1930s.[2] He saw **AC** as the result of a natural extrapolation of what he terms "extensional logic", valid in the realm of the finite, to infinite totalities. He considers formulation **AC3*** . In the special case in which A contains just two (or, more generally, finitely many elements), **AC3*** is essentially just the usual distributive law for $\wedge$ over $\vee$. Bernays now observes:

> *The universal statement of the principle of choice is then nothing other than the extension of an elementary-logical law* [i.e. the distributive law] *for conjunction and disjunction to infinite totalities, and the principle of choice constitutes thus a completion of the logical rules that concerns the universal and the existential judgment, that is, of the rules of existential inference,*

---

[1] In topos theory **AC** is treated precisely as a "postulate" in the Greek sense. For there the role of **AC** is to single out toposes of constant sets from general toposes of varying sets, in much the same way that the parallel postulate has come to be employed to single out flat geometries from curved ones. On the other hand, it is striking that - as pointed out above - in his 1908 formulation of **AC** Zermelo presents it as a genuine axiom, as opposed to the mere postulatory form in which it was presented in 1904.

[2] Bernays [1930-31].

whose application to infinite totalities also has the meaning that certain elementary laws for conjunction and disjunction are transferred to the infinite.

He goes on to remark that the principle of choice "is entitled to a special position only to the degree that the *concept of function* is required for its formulation." Most striking is his further assertion that the concept of function "in turn receives an adequate implicit characterization only through the principle of choice."

What Bernays seems to be saying is that in asserting the antecedent of **AC3***, in this case $\forall x \in A \exists y \in A \; R(x,y)$, one is implicitly asserting the existence of a function f: $A \to A$ for which $R(x,fx)$ holds for all x —that is, the consequent of **AC3***. On the surface, this seems remarkably similar to the justification of **AC** under the constructive interpretation of the quantifiers in which, let us remind ourselves once again, the assertability of an alternation of quantifiers $\forall x \exists y R(x,y)$ means *precisely* that one is given a function f for which $R(x,fx)$ holds for all x. However, Bernays goes on to draw the conclusion that, for the concept of function arising in this way, "the existence of a function with a [given] property in no way guarantees the existence of a concept-formation through which a determinate function with [that] property is uniquely fixed." In other words, the existence of a function may be asserted without the ability to provide it with an explicit definition[1]. This is incompatible with strict constructivism.

Bernays and the constructivists both affirm **AC3** through the claim that its antecedent and its consequent *have the same meaning*. But there is a difference, namely that, while Bernays in essence agrees with the constructive interpretation in treating the

[1] This fact, according to Bernays, renders the usual objections against the principle of choice invalid, since these latter are based on the misapprehension that the principle " claims the possibility of a choice".

quantifier block $\forall x \exists y$ as meaning $\exists f \forall x$, he interprets the existential quantifier in the latter *classically*, so that in affirming "there is a function " it is not necessary, as under the constructive interpretation, actually to be *given* such a function.

ZERMELO'S 1904 AND 1908 FORMULATIONS OF AC CONTRASTED TYPE-THEORETICALLY

We have seen that Zermelo's 1904 formulation of **AC**, in particular in its **AC3** version (more exactly, its **ACT** version) is provable in Constructive Type Theory. However, this is not the case for Zermelo's 1908 formulation, the combinatorial Axiom of Choice **CAC.** This was pointed out by Martin-Löf[1], who used a simplified form of Constructive Type Theory as a setting within which to contrast the two forms of **AC.** In Constructive Type Theory, according to Martin-Löf, the essential difference between these two forms of **AC** can be seen as arising from the implicit use of different realizations of the concept of set. The first and most basic, the intensional concept of set, is that of a plurality whose elements are taken to be equal when they are identical in the intensional sense of Constructive Type Theory. The second, the extensional concept of set, is that of a plurality whose elements are taken to be equal when they are "extensionally" equal in the usual set-theoretical sense. This amounts to taking an extensional set to be a(n) (intensional) set equipped with an equivalence relation representing the "extensional" equality of its elements, that is, a pair $\boldsymbol{S} = (S, =_S)$ where S is a set and $=_S$ is an equivalence relation on S. We shall use bold-face italic letters in this way to denote extensional sets.

[1] Martin-Löf [2006].

Because the formulation of the constructively provable version **AC3** of the Axiom of Choice involves just intensional sets in the above sense, it is natural to call **AC3** in this context the *intensional* Axiom of Choice. This, as we have seen, is provable within constructive type theory, and here we shall label it simply **AC.** Thus **AC** is the assertion that, for any sets I and S,

$$R \subseteq I \times S \wedge \forall x \in I \exists x \in S\, R(i, x) \Rightarrow (\exists f: I \to S) \forall x \in I\, R(i, fi)$$

We now want to formulate the corresponding choice principle for extensional sets. To do this we need to introduce the notions of extensional relation and extensional function. Thus let ***I*** and ***S*** be extensional sets. A relation R between I and S is called *extensional* , Ext(R), if it satisfies the conditions

$$i =_I j \Rightarrow [R(i, x) \Leftrightarrow R(j, x)] \quad x =_S y \Rightarrow [R(i, x) \Leftrightarrow R(i, y)]\,.$$

A function $f: I \to S$ is called *extensional* if $\forall i \in I \forall j \in I\, (i =_I j \Rightarrow fi =_S fj)$. We write $f: I \xrightarrow{ext} S$ to indicate that f is extensional. Then the Axiom of Choice for extensional sets takes the form:

ExtAC[1] $\quad R \subseteq I \times S \wedge \text{Ext}(R) \wedge \forall i \in I \exists x \in S\, R(i, x) \Rightarrow (\exists f: I \xrightarrow{ext} S)$

$$\forall i \in I\, R(i, fi).$$

Martin-Löf shows that, when suitably formulated in an "extensional" form, **CAC** is equivalent to **ExtAC** (as well as to some other principles) within his simplified version of Constructive Type Theory. That is, Zermelo's 1904 version of **AC**,

---

[1] **ExtAC** is to be distinguished from the extensional versions of the Axiom of Choice **UEAC** and **EAC** formulated within the weak set theory **WST** introduced in the previous chapter. In fact, formulated within **WST**, **ExtAC** is readily seen to be "in between" the two other versions in that the implications **UEAC** ⇒ **ExtAC** ⇒ **EAC** are provable in **WST.** Since, in **WST, EAC** ⇒ **REMS** (Thm. 2(c) of Ch. V), it also follows that **ExtAC** ⇒ **REMS.**

the "intensional" version, is constructively valid, but the 1908 version, the "extensional" version, is not[1].

We are going to present these arguments within the weak set theory **WST** introduced in the previous chapter. One difference between using Constructive Type Theory and **WST** as a background theory should be noted. In Constructive Type Theory, **AC** is provable, but in **WST,** it is not. Thus our arguments will be formulated within **WST + AC**.

In order to formulate a suitably "extensional" version of **CAC**, we need to introduce the concept of an extensional family of subsets of an extensional set. First, what should we take an extensional subset of an extensional set $(S, =_S)$ to be? Precisely what is called in set theory a subset *saturated* with respect to the equivalence relation $=_S$, that is, a subset $X \subseteq S$ satisfying $x =_S y \Rightarrow [x \in X \Leftrightarrow y \in X]$. Granted this, we make the following definition: given an extensional set $\mathbf{S} = (S, =_S)$, an *extensionally indexed family of disjoint extensional subsets of* **S** is specified by the following data:

- An extensional set $\boldsymbol{I} = (I, =_I)$
- A family $\{A_i: i \in I\}$ of subsets of S satisfying
  - (i) $x =_S y \Rightarrow [x \in A_i \Leftrightarrow y \in A_i]$
  - (ii) $i =_I j \Rightarrow A_i \approx A_i$ [2]
  - (iii) $\exists x[x \in A_i \cap A_j \Rightarrow i =_I j]$
  - (iv) $\forall i \in I \exists x \in S\ (x \in A_i)$

Let us abbreviate all this to EDF($\{A_i: i \in I\}$, **S**)**.** Then the extensional version of **CAC** may be written

---

[1] This exactly reverses Zermelo's essentially realist view of the matter. For he states that the 1904 version of **AC** was " somewhat tainted with subjectivity", and so presumably unacceptable, while in his eyes the "purely objective character" of the 1908 version "is immediately evident".

[2] Recall that $X \approx Y$ means that X and Y have the same elements, i.e. $\forall x[x \in X \Leftrightarrow x \in Y]$.

ExtCAC $\quad \mathrm{EIF}(\{A_i: i \in I\}, \mathbf{S}) \Rightarrow \exists S' \subseteq S\ [\forall x \forall y[x =_S y \Rightarrow [x \in S' \Leftrightarrow y \in S']] \wedge \forall i{\in}I \exists!^S x(x \in S' \cap A_i)]$

Here we have written $\exists!^S x$ for the "S-extensionalized" version of the unique existential quantifier $\exists!$: thus $\exists!^S x\varphi(x)$ is an abbreviation for $\exists x{\in}S\varphi(x) \wedge [\forall x{\in}S \forall y{\in}S[\varphi(x) \wedge \varphi(y) \Rightarrow x =_S y]]$.

Thus **ExtCAC** says that every extensionally disjoint family has an extensional choice set.

We also state the extensional version of **AC4.** For this we define a function $f\colon I \to S$ to be *extensionally epic*, written $f\colon \boldsymbol{I} \overset{ext}{\twoheadrightarrow} \mathbf{S}$, if f is extensional and $\forall x{\in}S \exists i{\in}I(f(i) =_S x)$. In its extensional version **AC4** takes the form

Epi $\quad f\colon S \overset{ext}{\twoheadrightarrow} \boldsymbol{I} \Rightarrow (\exists g\colon I \xrightarrow{ext} S) \forall i{\in}I\, f(g(i)) =_I i.$

Thus **Epi** says that every extensionally epic function has an extensional right inverse.

Finally we again recall **AC5** (*unique representatives can be picked from the equivalence classes of any given equivalence relation*):

AC5 $\quad Eq(R, I) \Rightarrow (\exists f\colon I \to I)\ [\forall i{\in}I\, R(i, fi) \wedge \forall i{\in}I\, \forall j{\in}I\, R(i, j) \Rightarrow fi = fj]$.

Now we can prove Martin-Löf's result in the form of the

Theorem. *In* **WST + AC,** *the principles* **ExtAC, ExtCAC, Epi,** *and* **AC5** *are all equivalent.*

Proof. We argue informally in In **WST + AC**

ExtAC ⇒ ExtCAC. Assuming $\mathrm{EDF}(\{A_i: i \in I\}, \mathbf{S})$, apply **ExtAC** to the relation $R(i, x) \equiv (x \in A_i)$ to get $f\colon I \xrightarrow{ext} S$ such that $\forall i{\in}I\, fi \in A_i$. Now define $S' = \{x{\in}S: \exists j{\in}I(x =_S fj)\}$ Then clearly $x =_S y \Rightarrow [x \in S' \Leftrightarrow y \in S']$, so it only remains to show that any pair of members of $S' \cap A_i$ are $=_S$ -equivalent. Suppose then that $x, y \in S' \cap A_i$. Then $x =_S fj$ and $y =_S fk$ for some $j, k \in I$. Now $x \in A_i$ and $x =_S fj$ gives $fj \in A_i$ by (i), and this, together with $fj \in A_j$ gives $i =_I j$

by (iii). Similarly $i =_I k$. Hence $j =_I k$. From the extensionality of f we deduce $fj =_S fk$, whence $x =_S y$. **ExtCAC** follows.

ExtCAC ⇒ Epi. Suppose $f: S \overset{ext}{\twoheadrightarrow} \mathbf{I}$. For $i \in I$ define $A_i = \{x \in S: fx =_I i\}$. It is then easily verified that EDF($\{A_i: i \in I\}$, **S**) holds. Applying **ExtCAC**, we get a subset $S' \subseteq S$ for which

(*) $[\forall x \forall y[x =_S y \Rightarrow [x \in S' \Leftrightarrow y \in S']] \wedge \forall i \in I \exists !^S x(x \in S' \cap A_i)]$.

Next, apply **IAC** to the relation $R(i, x) \equiv (x \in S' \cap A_i)$ to get $g: I \rightarrow S$ for which $\forall i \in I(gi \in S' \cap A_i)$. It follows that $\forall i \in I\, f(g(i)) =_I i$, so it only remains to show that g is extensional. Given $i, j \in I$, we have $gi \in S' \cap A_i$ and $gj \in S' \cap A_j$. So if $i =_I j$, then $A_i \approx A_j$, whence $gi \in S' \cap A_j$. But now from $gj \in S' \cap A_j$ and the second conjunct of (*) it follows that $gi =_S gj$. So g is extensional and **Epi** follows.

Epi ⇒ AC5. Let R be an equivalence relation on a set I and write Id_I for the identity relation on I. Then clearly the identity map on I is extensionally epic from (I, Id_I) to (I, R). Assuming **Epi**, there is then a function $f: I \rightarrow I$ for which $R(gi, i)$ and $R(i, j) \Rightarrow Id_I(fi, fj)\,) \Rightarrow fi = fj$. This gives **AC5.**

AC5 ⇒ ExtAC. Assume the antecedent of **ExtAC,** and use **IAC** to obtain a choice function $f: I \rightarrow S$ for which $\forall i \in I(R(i, fi)$. Assuming **AC5,** get a $g: I \rightarrow I$ satisfying $gi =_I i$ and $i =_I j \Rightarrow gi = gj$. Let $h = f \circ g$. Then h is extensional, since $i =_I j \Rightarrow gi = gj \Rightarrow fgi =_S fgj$. Also, for $i \in I$, we have $R(gi, fgi)$, i.e. $R(gi, hi)$. Since $gi =_I i$ and R is extensional we conclude that $R(gi, hi)$. **ExtAC** follows. ■

INTENSIONAL AND EXTENSIONAL AC COMPARED

We have noted that, from a set-theoretic point of view, the affirmability of the intensional Axiom of Choice in Constructive Type Theory corresponds to the fact that, for any doubly-indexed family of sets $\{A_{ij} : i \in I, j \in J\}$ there is a bijection

(1) $$\prod_{i \in I} \coprod_{j \in J} A_{ij} \cong \coprod_{f \in J^I} \prod_{i \in I} A_{if(i)} .$$

This bijection is easily described: to wit, it is the map

(2) $$g \mapsto (\pi_1 \circ g, \pi_2 \circ g) = g^*,$$

where π_1, π_2 are the projections of ordered pairs onto their first and second coordinates.

Note that

(3) for $g \in \prod_{i \in I} \coprod_{j \in J} A_{ij}$, g^* is a pair of functions (e, f) with $f \in J^I$ and $e \in \prod_{i \in I} A_{if(i)}$.

We have also observed that in set theory, the Axiom of Choice is equivalent to the assertion that, for any doubly-indexed family of sets $\{A_{ij} : i \in I, j \in J\}$,

(4) $$\prod_{i \in I} \bigcup_{j \in J} A_{ij} \subseteq \bigcup_{f \in J^I} \prod_{i \in I} A_{if(i)} .$$

Let us attempt to elucidate, within set theory, the connection between the two formulations **of AC** given by (1) and (4).

First observe that there is a natural epic map

$$\prod_{i \in I} \coprod_{j \in J} A_{ij} \twoheadrightarrow \prod_{i \in I} \bigcup_{j \in J} A_{ij}$$

given by

$$g \mapsto \pi_1 \circ g$$

Now let us assume that this map has a right inverse u, that is,

$$u\colon \prod_{i \in I} \bigcup_{j \in J} A_{ij} \to \prod_{i \in I} \coprod_{j \in J} A_{ij}$$

satisfies

(5) $$\pi_1 \circ u(k) = k,$$

for all $k \in \prod_{i \in I} \bigcup_{j \in J} A_{ij}$.

We are now in a position to use (1), together with the existence of the map u, to obtain (4). Given any $k \in \prod_{i \in I} \bigcup_{j \in J} A_{ij}$, under the natural bijection (2), $u(k)$ is correlated with the pair of maps

$$(\pi_1 \circ u(k)\,,\, \pi_2 \circ u(k)),$$

i.e., using (5), with

$$(k\,,\, \pi_2 \circ u(k)).$$

Writing $f = \pi_2 \circ u(k)$, it follows from (3) that

$$f \in J^I \text{ and } k \in \prod_{i \in I} A_{if(i)},$$

whence

$$k \in \bigcup_{f \in J^I} \prod_{i \in I} A_{if(i)}.$$

Thus we have derived (4).

Of course, from a formal standpoint the argument we have given amounts merely to a derivation in set theory of (4) from **AC4**, using the set-theoretically provable principle (1) as a step along the way. However, this can be put in much more suggestive terms. For each $g \in \prod_{i \in I} \coprod_{j \in J} A_{ij}$ and each $i \in I$, the identity of the (unique) $j \in J$ for which $g(i) \in A_{ij}$. is, as it were, information "coded" into $\prod_{i \in I} \coprod_{j \in J} A_{ij}$. To apply the epi $\prod_{i \in I} \coprod_{j \in J} A_{ij} \twoheadrightarrow \prod_{i \in I} \bigcup_{j \in J} A_{ij}$ is, thus, in effect, to discard this information: after the application, one only "knows" that $g(i)$ is a member of some A_{ij} but not precisely which. The map u furnished by **AC4** essentially resupplies that information. So starting with $k \in \prod_{i \in I} \bigcup_{j \in J} A_{ij}$, if one applies u to it, and then applies to the result the bijection (2), one winds up with a map $f \in J^I$ for which $k(i) \in A_{if(i)}$ for all $i \in I$. This is exactly what is demanded by (1).

A LAST LOOK AT AC AND THE PROPOSITIONS-AS-TYPES DOCTRINE

As we have seen, under the propositions- as-types interpretation, **AC** is provable, and so *a fortiori* has no "untoward" logical consequences within that framework. On the other hand, we also know that in intuitionistic set theory, or in the internal language of a topos this is far from being the case, for, as Diaconescu's theorem shows, in the latter **AC** implies **LEM**. This prompts the question: what modification needs to be made to the propositions-as-types doctrine so as to yield the set- or topos-theoretic interpretation of **AC**? An illuminating answer to this question has been given by Maietti [2005] through the use of so-called *monotypes* (or mono-objects), that is, (dependent) types containing at most one entity or having at most one proof. In $\mathcal{S}et$, mono objects are *singletons*, that is, sets containing at most one element.

Monotypes correspond to monic maps. This can be illustrated concretely by considering the toposes $\mathcal{B}iv$ of bivariant sets introduced above and the topos $\mathcal{I}ndset$ of *indexed* sets. The objects of $\mathcal{I}ndset$ are indexed sets of the form $\mathfrak{M} = \{<i, M_i>: i \in I\}$ with arrows $f: \mathfrak{M} \to \mathfrak{N}$ indexed sets of maps $f_i: M_i \to N_i$ .It can be shown that these two categories are equivalent. If we think of (the objects of) $\mathcal{S}et$ as representing simple or static types, then (the objects of) $\mathcal{I}ndset$, and hence also of $\mathcal{B}iv$, represent dependent or variable types. It is easily seen that a monotype, or object, in $\mathcal{I}ndset$, is precisely an object M for which each M_i has at most one element. Moreover, under the equivalence between $\mathcal{I}ndset$ and $\mathcal{B}iv$, such an object corresponds to a monic map- object in $\mathcal{B}iv$.

Now consider $\mathcal{B}iv$ as a topos. Under the topos-theoretic interpretation in $\mathcal{B}iv$, formulas correspond to monic arrows, which in turn correspond to mono-objects in $\mathcal{I}ndset$. Carrying these correspondences over entirely to *Indset* yields the sought modification of the propositions-as-types interpretation to bring it into line with the topos-theoretic interpretation of formulas, namely, to take formulas or propositions to correspond to *mono-objects*, rather than to *arbitrary* objects. Let us call this the *formulas-as-monotypes* interpretation.

Finally let us reconsider **AC** under the formulas-as-monotypes interpretation within *Set*. It will be convenient to rephrase **AC** as the assertion

(*) $$\forall i \in I\, \exists j \in J\, M_{ij} \Leftrightarrow \exists f \in J^I\, \forall i \in I\, M_{if(i)}$$

where $<M_{ij}: i \in I, j \in J>$ is any doubly indexed family of propositions (or sets). In the propositions-as-types interpretation, (*) corresponds to the existence of an isomorphism between $\prod_{i \in I} \coprod_{j \in J} M_{ij}$ and $\coprod_{f \in J^I} \prod_{i \in I} M_{if(i)}$. On the other hand, **AC** interpreted in the usual way, that is, using the rules of topos semantics, can be presented in the form of the distributive law

(**) $$\bigcap_{i \in I} \bigcup_{j \in J} M_{ij} = \bigcup_{f \in J^I} \bigcap_{i \in I} M_{if(i)}.$$

In the propositions-as-types interpretation (as applied to *Set*), the universal quantifier $\forall i \in I$ corresponds to the product $\prod_{i \in I}$ and the existential quantifier $\exists i \in I$ to the coproduct, or disjoint sum, $\coprod_{i \in I}$. Now in the formulas-as-monotypes interpretation, wherein formulas correspond to singletons, $\forall i \in I$ continues to correspond to $\prod_{i \in I}$, since the product of singletons is still a singleton. But the interpretation of $\exists i \in I$ is changed. In fact, the interpretation of $\exists i \in I\, M_i$ (with each M_i a singleton) now becomes $[\coprod_{i\in I} M_i]$, where for each set X, $[X] = \{u: u = 0 \wedge \exists x.\ x \in X\}$ is the *canonical singleton* associated with X.

It follows that, under the formulas-as-monotypes interpretation, the proposition $\forall i\in I\ \exists j\in J\ M_{ij}$ is interpreted as the singleton

(1) $$\prod_{i\in I}[\coprod_{j\in J} M_{ij}]$$

and the proposition $\exists f\in J^I\ \forall i\in I\ M_{if(i)}$ as the singleton

(2) $$[\coprod_{f\in J^I}\prod_{i\in I} M_{if(i)}].$$

Under the formulas-as-monotypes interpretation **AC** would be construed as asserting the existence of an isomorphism between (1) and (2).

Now it is readily seen that to give an element of (1) amounts to no more than affirming that, for every $i \in I$, $\bigcup_{j\in J} M_{ij}$ is nonempty. But to give an element of (2) amounts to specifying maps $f \in J^I$ and g with domain I such that $\forall i\in I\ g(i) \in M_{if(i)}$. It follows that to assert the existence of an isomorphism between (1) and (2), that is, to assert **AC** under the formulas-as-monotypes interpretation, is tantamount to asserting **AC** in the form (**), so leading in turn to classical logic. This is in sharp contrast with **AC** under the propositions-as-types interpretation, under which, let us reiterate, its assertion is automatically correct and so has no nonconstructive consequences.

Appendix I
Intuitionistic Logic

(*Free*) *intuitionistic first-order logic* has the following axioms and rules of inference.

Axioms

$$\alpha \to (\beta \to \alpha)$$
$$[\alpha \to (\beta \to \gamma) \to [(\alpha \to \beta) \to (\alpha \to \gamma)]$$
$$\alpha \to (\beta \to \alpha \wedge \beta)$$
$$\alpha \wedge \beta \to \alpha \qquad \alpha \wedge \beta \to \beta$$
$$\alpha \to \alpha \vee \beta \qquad \beta \to \alpha \vee \beta$$
$$[\alpha \to (\beta \to \gamma) \to [(\alpha \to \beta) \to (\alpha \to \gamma)]$$
$$(\alpha \to \gamma) \to [(\beta \to \gamma) \to (\alpha \vee \beta \to \gamma)]$$
$$(\alpha \to \beta) \to [(\alpha \to \neg\beta) \to \neg\alpha]$$
$$\neg\alpha \to (\alpha \to \beta)$$
$$\alpha(t) \to \exists x\alpha(x) \qquad \forall x\alpha(x) \to \alpha(y)$$
(x free in α and t free for x in α)
$$x = x \qquad \alpha(x) \wedge x = y \to \alpha(y)$$

Rules of Inference

$$\frac{\alpha,\ \alpha \to \beta}{\beta}$$
(all free variables of α free in β)

$$\frac{\beta \to \alpha(x)}{\beta \to \forall x\alpha(x)} \qquad \frac{\alpha(x) \to \beta}{\exists x\alpha(x) \to \beta}$$
(x not free in β)

Classical first-order logic is obtained by adding to the intuitionistic system the rule of inference

$$\frac{\neg\neg\alpha}{\alpha}$$

In intuitionistic logic none of the classically valid logical schemes

LEM (Law of Excluded Middle) $\alpha \vee \neg\alpha$

LDN (Law of Double Negation) $\neg\neg\alpha \rightarrow \alpha$

DEM (de Morgan's Law) $\neg(\alpha \wedge \beta) \rightarrow \neg\alpha \vee \neg\beta$

are derivable. However **LEM** and **LDN** are intuitionistically equivalent and **DEM** is intuitionistically equivalent to the Weakened Law of Excluded Middle:

WLEM $\neg\alpha \vee \neg\neg\alpha$.

Also the weakened form of **LDN** for negated statements,

WLDN $\neg\neg\neg\alpha \rightarrow \neg\alpha$

is intuitionistically derivable. It follows that any formula intuitionistically equivalent to a negated formula satisfies **LDN**.

Appendix II

Basic Concepts of Category Theory

A *category* $\mathscr{C}$ is determined by first specifying two classes $Ob(\mathscr{C})$, $Arr(\mathscr{C})$—the collections of *$\mathscr{C}$-objects* and *$\mathscr{C}$-arrows* (or *morphisms*). These collections are subject to the following axioms:

- Each $\mathscr{C}$-arrow f is assigned a pair of $\mathscr{C}$-objects dom(f), cod(f) called the *domain* and *codomain* of f, respectively. To indicate the fact that $\mathscr{C}$-objects X and Y are respectively the domain and codomain of f we write $f: X \to Y$ or $X \xrightarrow{f} Y$. The collection of $\mathscr{C}$-arrows with domain X and codomain Y is written $\mathscr{C}(X, Y)$.
- Each $\mathscr{C}$-object X is assigned a $\mathscr{C}$-arrow $1_X: X \to X$ called the *identity arrow* on X. (1_X is sometimes written *id*.)
- Each pair f, g of $\mathscr{C}$-arrows such that cod(f) = dom(g) is assigned an arrow $g \circ f$: dom(f) $\to$ cod(g) called the *composite* of f and g. Thus if $f: X \to Y$ and $g: Y \to Z$ then $g \circ f: X \to Z$. We also write $X \xrightarrow{f} Y \xrightarrow{g} Z$ or gf for $g \circ f$. Arrows f, g satisfying cod(f) = dom(g) are called *composable.*
- *Associativity law.* For composable arrows (f, g) and (g, h), we have $h \circ (g \circ f) = (h \circ g) \circ f$.
- *Identity law.* For any arrow $f: X \to Y$, we have $f \circ 1_X = f = 1_Y \circ f$.

As a fundamental example of a category, we have the category $\mathcal{S}et$ of sets whose objects are all sets and whose arrows are all maps between sets (strictly, triples (f, A, B) with domain$(f) = A$ and range$(f) \subseteq B$.) Other examples of categories are the category $\mathcal{G}rp$ of groups, with objects all groups and arrows all group homomorphisms and the category $\mathcal{T}op$ of topological spaces with objects all topological spaces and arrows all continuous maps. A category with just one object may be identified with a *monoid,* that is, algebraic structures with an associative multiplication and an identity element. At the other extreme, a category in which there is at most one arrow between any pair of objects may be identified with a *preordered class,* that is, a class equipped with a reflexive transitive relation.

A *subcategory* $\mathcal{C}$ of a category $\mathcal{D}$ is any category whose class of objects and arrows is included in the class of objects and arrows of $\mathcal{D}$, respectively, and which is closed under domain, codomain, identities, and composition. If, further, for any $\mathcal{C}$-objects C, C' we have $\mathcal{C}(C, C') = \mathcal{D}(C, C')$, we shall say that $\mathcal{C}$ is a *full* subcategory of $\mathcal{D}$.

BASIC CATEGORY-THEORETIC DEFINITIONS

Commutative diagram	Diagram of objects and arrows such that the arrow obtained by composing the arrows of any connected path depends only on the endpoints of the path.
Initial object	Object 0 suach that, for any object X, there is a unique arrow $0 \rightarrow X$. (In $\mathcal{S}et$, 0 is $\varnothing$.)
Terminal object	Object 1 such that, for any object X, there is a unique arrow $X \rightarrow 1$. (In $\mathcal{S}et$, 1 is $\{\varnothing\}$.)
Element of an object X	Arrow $1 \rightarrow X$.

Monic arrow $X \rightarrowtail Y$	Arrow f: $X \to Y$ such that, for any arrows g, h: $Z \to X$, $fg = fh \Rightarrow g = h$. (In $\mathcal{S}et$, injective map.)
Epic arrow $X \twoheadrightarrow Y$	Arrow f: $X \to Y$ such that, for any arrows g, h: $Y \to Z$, $gf = hf \Rightarrow g = h$. (In $\mathcal{S}et$, surjective map.)
Isomorphism $X \cong Y$	Arrow f: $X \to Y$ for which there exists g: $Y \to X$ such that $gf = 1_X$, $fg = 1_Y$. (In $\mathcal{S}et$, bijective map.)
Product of objects X, Y	Object $X \times Y$ with arrows (*projections*) $X \xleftarrow{\pi_1} X \times Y \xrightarrow{\pi_2} Y$ such that any diagram 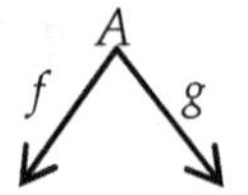 can be uniquely completed to a commutative diagram 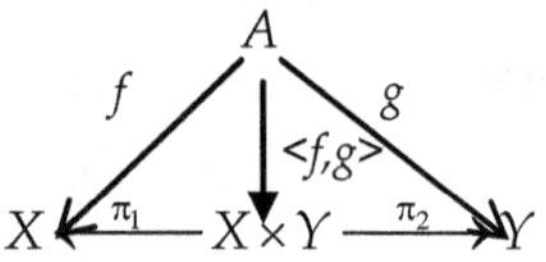 In $\mathcal{S}et$, $X \times Y$ is the usual Cartesian product of X and Y.
Product of arrow f_1: $X_1 \to Y_1$, f_2: $X_2 \to Y_2$	The arrow $f_1 \times f_2 = \langle f_1\pi_1, f_2\pi_2\rangle$: $X_1 \times Y_1 \to X_2 \times Y_2$
Diagonal arrow on object X	Unique arrow δ_X: $X \to X \times X$ making the diagram 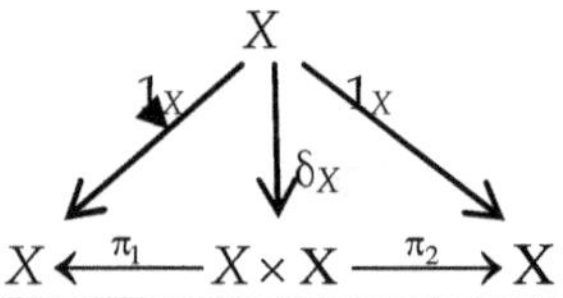 commute.
Coproduct of objects X, Y	Object $X + Y$ with arrows (*injections*) $X \xrightarrow{\sigma_1} X + Y \xleftarrow{\sigma_2} Y$ such that any diagram 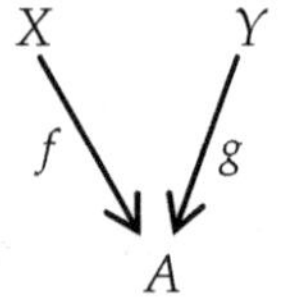 can be uniquely completed to a commutative diagram 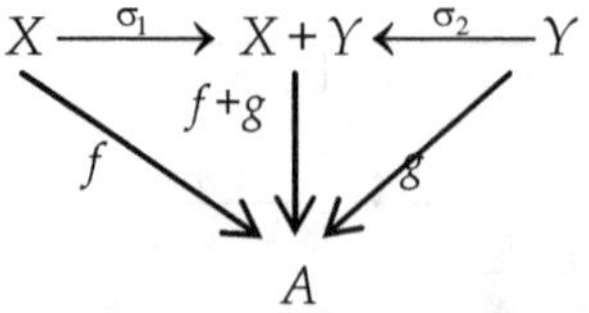

	In $\mathcal{S}et$, $X + Y$ is the disjoint union of X and Y.
Pullback diagram	Commutative diagram of the form $\begin{array}{ccc} A & \longrightarrow & B \\ \downarrow & & \downarrow f \\ C & \xrightarrow{g} & D \end{array}$ such that for any commutative diagram $\begin{array}{ccc} X & \longrightarrow & B \\ \downarrow & & \downarrow f \\ C & \xrightarrow{g} & D \end{array}$ there is a unique $X \xrightarrow{!} A$ such that $\begin{array}{cccc} X & & & \\ & \searrow & & \searrow \\ & A & \longrightarrow & B \\ & \downarrow & & \downarrow f \\ & C & \xrightarrow{g} & D \end{array}$ (with $X \to C$) commutes.
Equalizer of pair of arrows $A \overset{f}{\underset{g}{\rightrightarrows}} B$	Arrow $E \xrightarrow{e} A$ such that $fe = ge$ and, for any arrow $E' \xrightarrow{e'} A$ such that $fe' = ge'$ there is a unique $E' \xrightarrow{u} E$ such that $eu = e'$.
Truth value object or *subobject classifier*	Object Ω together with arrow $\top: 1 \to \Omega$ such that every monic $m: A \rightarrowtail B$ can be uniquely extended to a pullback diagram of the form $\begin{array}{ccc} A & \longrightarrow & 1 \\ m \downarrow & & \downarrow \top \\ B & \xrightarrow{\chi(m)} & \Omega \end{array}$ and conversely every diagram of the form $\begin{array}{ccc} & & 1 \\ & & \downarrow \top \\ B & \longrightarrow & \Omega \end{array}$ has a pullback. $\chi(m)$ is called the *characteristic*

	arrow of *m*. The *maximal* characteristic arrow T_A, or simply *T*, on *A*, is defined to be the characteristic arrow of 1_A. The characteristic arrow of $0 \rightarrowtail 1$ is written $\perp: 1 \to \Omega$. (In *Set*, Ω is the set $2 = \{0, 1\}$ and $\perp$ is the map $1 \to 2$ taking value 1.)
Power object of an object *X*	Object *PX* together with arrow $e_X: X \times PX \to \Omega$ such that, for any $f: X \times Y \to \Omega$, there is a unique $f^*: Y \to PX$ such that 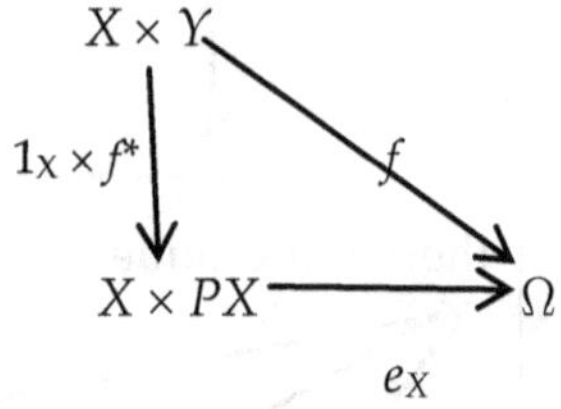 commutes. (In *Set*, *PX* is the power set of *X* and e_X is the characteristic function of the membership relation between *X* and *PX*.)
Exponential of objects *Y*, *X*	Object Y^X together with arrow $ev: X \times Y^X \to Y$ such that, for any $f: X \times Z \to Y$, there is a unique $f^*: Z \to Y^X$ such that 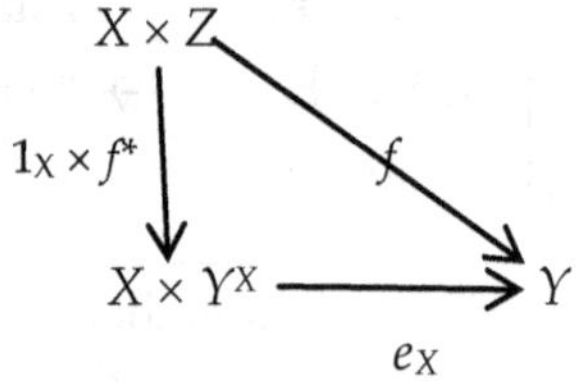 commutes. (In *Set*, Y^X is the set of all maps $X \to Y$ and *ev* is the map sending (x, f) to $f(x)$.)
Product of indexed set of objects $\{A_i: i \in I\}$	Object $\prod_{i \in I} A_i$ together with arrows $\pi_i: \prod_{i \in I} A_i \to A_i$ $(i \in I)$ such that, for any arrows $f_i: B \to A_i$ $(i \in I)$ there is a unique arrow $h: B \to \prod_{i \in I} A_i$ such that, for each $i \in I$, the diagram

	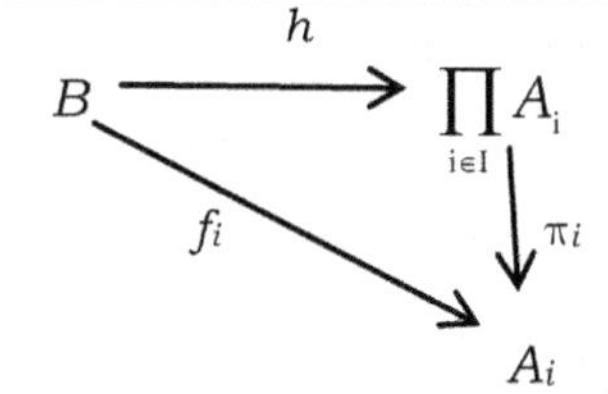 commutes. (In $\mathcal{S}et$, $\prod_{i\in I} A_i$ is the Cartesian product of the A_i and the π_I are projection maps.)
Coproduct of indexed set of objects $\{A_i: i \in I\}$	Object $\coprod_{i\in I} A_i$ together with arrows σ_i: $A_i \to \coprod_{i\in I} A_i$ $(i \in I)$ such that, for any arrows f_i: $A_i \to B$ $(i \in I)$ there is a unique arrow h: $\coprod_{i\in I} A_i \to B$ such that, for each $i \in I$, the diagram 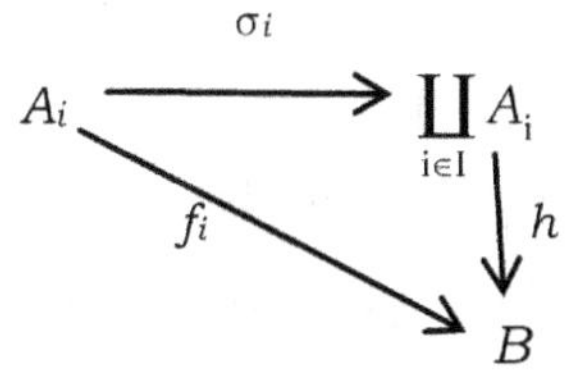 commutes. If each A_i is a fixed object A, $\coprod_{i\in I} A_i$ is called the *I-indexed copower of A*. (In $\mathcal{S}et$, $\coprod_{i\in I} A_i$ is the disjoint union of the A_i, i.e, the set $\bigcup_{i\in I} A_i \times \{i\}$. In particular the I-indexed copower of 1 in $\mathcal{S}et$ is the set $\{<\emptyset, i>: i \in I\}$.

A category is *cartesian closed* if it has a terminal object, as well as products and exponentials of arbitrary pairs of its objects. It is *finitely complete* if it has a terminal object, products of arbitrary pairs of its objects, and equalizers. A *topos* is a category possessing a terminal object 1, products, a truth-value object, and power objects. In particular $\mathcal{S}et$ is a topos. It can be shown that every topos has an initial object 0, is cartesian closed, finitely complete, and has coproducts of arbitrary pairs of its objects. A topos in which $0 \not\cong 1$ is said to be *nondegenerate.* A topos $\mathcal{E}$ is *well-copowered* if arbitrary set-indexed copowers of 1 exist in $\mathcal{E}$. In particular $\mathcal{Set}$ is well-copowered.

More on products in a category. A *product* of objects $A_1, ..., A_n$ in a category $\mathcal{C}$ is an object $A_1 \times ... \times A_n$ together with arrows π_i: $A_1 \times ... \times A_n \to A_i$ for $i = 1, ..., n$, such that, for any arrows f_i: $B \to A_i$, $i = 1, ..., n$, there is a unique arrow , denoted by $<f_1, ..., f_n>$: $B \to A_1 \times ... \times A_n$ such that $\pi_i \circ <f_1, ..., f_n> = f_i$, $i = 1, ..., n$. Note that, when $n = 0$, $A_1 \times ... \times A_n$ is the terminal object 1. The category is said to *have finite products* if $A_1 \times ... \times A_n$ exists for all $A_1, ..., A_n$. If $\mathcal{C}$ has binary products, it has finite products, since we may take $A_1 \times ... \times A_n$ to be $A_1 \times (A_2 \times (... \times A_n)...)$. It is easily seen that the product operation is, up to isomorphism, commutative and associative. The relevant isomorphisms are called *canonical* isomorphisms.

A *functor* F: $\mathcal{C} \to \mathcal{D}$ between two categories $\mathcal{C}$ and $\mathcal{D}$ is a map that "preserves commutative diagrams", that is, assigns to each $\mathcal{C}$-object A a $\mathcal{D}$-object FA and to each $\mathcal{C}$-arrow f: $A \to B$ a $\mathcal{D}$-arrow Ff: $FA \to FB$ in such a way that, for any object A, $F(1_A) = 1_{FA}$ and, for any composable arrows f, g, we have $F(g \circ f) = Fg \circ Ff$.

A functor F: $\mathcal{C} \to \mathcal{D}$ is an *equivalence* if it is "an isomorphism up to isomorphism", that is, if it is

- *faithful*: $Ff = Fg \Rightarrow f = g$.
- *full*: for any h: $FA \to FB$ there is f: $A \to B$ such that $h = Ff$.
- *dense*: for any $\mathcal{D}$-object B there is a $\mathcal{C}$-object A such that $B \cong FA$.

Two categories are *equivalent,* written $\simeq$, if there is an equivalence between them. Equivalence is the appropriate notion of "identity of form" for categories.

Given functors F, G: $\mathscr{C} \to \mathscr{D}$, a *natural transformation* between F and G is a map η from the objects of $\mathscr{C}$ to the arrows of $\mathscr{D}$ satisfying the following conditions:

For each $\mathscr{C}$-object A, ηA is a $\mathscr{D}$-arrow $FA \to GA$;

for each $\mathscr{C}$-arrow f: $A \to A'$ the diagram

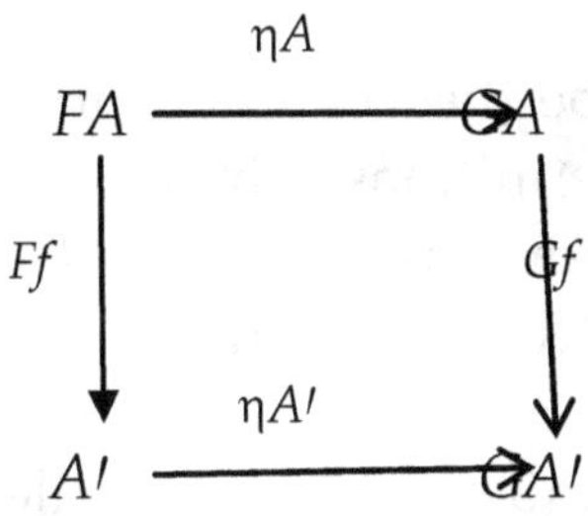

commutes.

Finally, two functors F: $\mathscr{C} \to \mathscr{D}$ and G: $\mathscr{D} \to \mathscr{C}$ are said to be *adjoint* to one another, written $F \dashv G$, if, for any objects A of $\mathscr{C}$, B of $\mathscr{D}$, there is a "natural" bijection between arrows $A \to GB$ in $\mathscr{C}$ and arrows $FA \to B$ in $\mathscr{D}$. To be precise, for each such pair A, B we must be given a bijection φ_{AB}: $\mathscr{C}(A, GB) \to \mathscr{D}(FA, B)$ satisfying the "naturality" conditions

- for each f: $A \to A'$ and h: $A' \to GB$, $\varphi_{AB}(h \circ f) = \varphi_{A'B}(h) \circ Ff$
- for each g: $B \to B'$ and h: $A \to GB'$, $\varphi_{AB}(Gg \circ h) = g \circ \varphi_{AB'}(h)$.

Under these conditions F is said to be *left adjoint* to G, and G *right adjoint* to F.

Bibliography

Aczel, P. [1978] The type-theoretic interpretation of constructive set theory. In A. MacIntyre, L. Pacholski, and J. Paris. eds., *Logic Colloquium 77.*, pp. 55-66. North-Holland.

Aczel, P. [1982]. The type-theoretic interpretation of constructive set theory: choice principles. In A. S. Troelstra and D. van Dalen, eds., *The L.E.J. Brouwer Centenary Symposium,* pp. 1-40. North-Holland.

Aczel, P. [1986]. The type-theoretic interpretation of constructive set theory: inductive definitions. In R. Barcan Marcus, G.J.W. Dorn and P. Weinegartner, eds., *Logic, Methodology and Philosophy of Science VII,* pp. 17-49. North-Holland.

Aczel, P. and N. Gambino [2002)] Collection principles in dependent type theory. In P. Callaghan, Z. Luo, J. McKinna and R. Pollack, eds. *Types for Proofs and Programs,* vol. 2277 of *Lecture Notes on Computer Science,* pp. 1-23. Springer.

Aczel, P. and N. Gambino [2005]. The generalized type-theoretic interpretation of constructive set theory. Manuscript available on first author's webpage www.cs.man.ac.uk/~petera/papers

Aczel, P. and M. Rathjen [2001]. Notes on Constructive Set Theory. Technical Report 40, Mittag-Leffler Institute, The Swedish Royal Academy of Sciences. Available on first author's webpage www.cs.man.ac.uk/~petera/papers

Alas, O.T. [1969] The axiom of choice and two particular forms of Tychonoff theorem. Portugal. Math. **28**, 75-6.

Balbes, R., and Dwinger, P. [1974] *Distributive Lattices.* University of Missouri Press, 1974.

Banach, S. and Tarski, A. [1924] . Sur la décomposition des ensembles de points en parties respectivement congruentes. *Fundamenta Mathematicae* **6**, 244-277.

Banaschewski, B. [1980] The duality of distributive continuous lattices. *Canadian J. of Math.* **32**, 385-94.

Banaschewski, B., and Bhutani, K., [1986] Boolean algebras in a localic topos. *Math. Proc. Camb. Phil. Soc.* **100** , 43-55.

Banaschewski, B. and Mulvey, C. [1980] Stone-Čech compactification of locales. *Houston Journal of Mathematics* **6**, 301-12.

Beeson, M.J. [1985] *Foundations of Constructive Mathematics.* Berlin: Springer-Verlag.

Bell, J. L. [1983] On the strength of the Sikorski extension theorem for Boolean algebras. *J. Symbolic Logic* **48**, 841-846.

Bell, J. L. [1988] *Toposes and Local Set Theories: An Introduction.* Clarendon Press, Oxford, 1988. Dover reprint 2007.

Bell, J. L. [1988a]. Some propositions equivalent to the Sikorski extension theorem for Boolean algebras. *Fundamenta Mathematicae* **130**, 51-5.

Bell, J. L. [1993] Hilbert's epsilon-operator and classical logic. *Journal of Philosophical Logic*, **22** .

Bell, J. L. [1993a] Hilbert's epsilon operator in intuitionistic type theories. *Math. Logic Quarterly*, **39.**

Bell, J. L. [1997]. Zorn's lemma and complete Boolean algebras in intuitionistic type theories. *J. Symbolic Logic.* **62**, 1265-1279.

Bell, J. L. [1999]. Boolean algebras and distributive lattices treated constructively. *Math. Logic Quart.* **45**, 135-143.

Bell, J. L. [2003]. Some new intuitionistic equivalents of Zorn's Lemma. *Arch. Math. Logik* **42**, 811-814.

Bell, J. L. [2005]. *Set Theory: Boolean-valued Models and Independence Proofs*. Clarendon Press, Oxford.

Bell, J. L. [2006]. Choice principles in intuitionistic set theory. In *A Logical Approach to Philosophy*. Springer.

Bell, J.L. [2006a] Abstract and Variable Sets in Category Theory. In *What is Category Theory?* Polimetrica.

Bell, J. L. [2008] The axiom of choice and the law of excluded middle in weak set theories. *Mathematical Logic Quarterly*. **54,** no. 2., 194-201.

Bell, J.L. and Fremlin, D. [1972] The maximal ideal theorem for lattices of sets. *Bull. London Math. Soc.* **4**, 1-2.

Bell, J. L. and Fremlin, D. [1972a] A geometric form of the axiom of choice. *Fundamenta Mathematicae* **77**, 167-170.

Bell, J. L. and Machover, M. [1977]. *A Course in Mathematical Logic.* North-Holland.

Bell, J. L. and Slomson, A.B. [2006]. *Models and Ultraproducts: An Introduction.* Dover.

Bernays, P. [1930-31]. Die Philosophie der Mathematik und die Hilbertsche Beweistheorie. Blätter für deutsche Philosophie **4**, pp. 326-67. Translated in Mancosu, *From Brouwer to Hilbert,* Oxford University Press, 1998.

Bernays, P. [1942]. A system of axiomatic set theory, Part III. *Journal of Symbolic Logic* **7**, 65-89.

Bénabou, J. (1958) Treillis locaux et paratopologies. *Séminaire Ehresmann* (Topologie et Géométrie Différentielle), 1re année (1957-8), exposé **2.**

Bishop, E. [1967] *Foundations of Constructive Analysis.* McGraw-Hill.

Bishop, E. and Bridges, D. [1985]. *Constructive Analysis.* Berlin: Springer.

Blass, A. [1977]. A model without ultrafilters. *Bull. d'Acad. Pol. des Sci.* **25**, 329-331.

Blass, A. [1979]. Injectivity, projectivity and the axiom of choice. *Trans. A.M.S.* 25, 329-331.

Blass, A. [1984]. Existence of bases implies the axiom of choice. In *Axiomatic Set Theory,* Baumgartner, Martin and Shelah, eds. Contemporary Mathematics Series, Vol. 31, American Mathematical Society, pp. 31-33.

Bourbaki, N. [1939]. *Eléments de mathématique. Premiere partie: Les structures fondementales de l'analyse. Livre I: Théorie des ensembles.* Paris: Hermann.

Bourbaki, N. [1950]. Sur le théorème de Zorn. *Arkiv der Matematik* **2**, 434-437.

Bourbaki, N. [1963] *Eléments de mathématique. Premiere partie: Les structures fondementales de l'analyse. Livre I: Théorie des ensembles. Seconde édition.* Paris: Hermann.

Cohen, P. J. [1963] *The independence of the axiom of choice.* Mimeographed.

Cohen, P.J. [1963a] The independence of the continuum hypothesis I. *Proceedings of the U.S. National Academy of Sciemces* **50,** 1143-48.

Cohen, P.J. [1964] The independence of the continuum hypothesis II. *Proceedings of the U.S. National Academy of Sciemces* **51,** 105-110.

Curry, H.B. and R. Feys [1958]. *Combinatory Logic.* North Holland.

Devidi, D. [2004]. Choice principles and constructive logics. *Philosophia Mathematica* (3), **12**, 222-243.

Diaconescu, R. [1975] Axiom of choice and complementation. *Proc. Amer. Math. Soc.* **51**, 176–8.

Dowker, C.H. and D. Papert (Strauss) [1966] Quotient frames and subspaces. *Proc. Lond. Math. Soc.* **16**, 275-296.

Dowker, C.H. and D. Papert (Strauss) [1966] [1972] Separation axioms for frames. *Colloq. Math. Socc. Janos Bolyai* **8,** 223-240.

Ehresmann, C. [1957] Gattungen von lokalen Strukturen. *Jber. Deutsch. Math.-Verein* **60**, 59-77.

Fraenkel, A. [1922] Zu den Grundlagen der Cantor-Zermeloschen Mengenlehre. *Mathematische Annalen* **86,** 230-237.

Fraenkel, A. [1922a] Über den Begriff 'definit' und die Unabhängigkeit des Auswahlsaxioms. *Sitzungsberichte der Preussischen Akademie der Wissenschaften, Physik-math. Klasse,* 253-257. Translated in Translated in van Heijenoort, From Frege to Gödel: A Source Book in Mathematical Logic 1879-1931, Harvard University Press, 1967, pp. 284-289.

Fraenkel, A. [1976] *Abstract Set Theory,* 4th Revised Edition. North-Holland.

Fraenkel, A., Y. Bar-Hillel and A. Levy [1973]. *Foundations of Set Theory,* 2nd edition. North-Holland.

Gandy, R.O [1956, 1959]. On the axiom of extensionality, Part I, *Journal of Symbolic Logic* **21,** 1956, pp. 36-48; Part II, *ibid.,* **24,** 1959, pp. 287-300.

Gelfand, I.M. [1939]. On normed rings. *Dokl. Akad. Nauk. USSR* **23** 430- 2.

Gelfand, I.M. [1941]. Normierte Ringe. *Mat. Sb.* **9 (51),** 3-24.

Gelfand, I. M. and A.N Kolmogorov. [1939]. On rings of continuous functions on topological spaces. *Dokl. Akad. Nauk. USSR* **22,** 11-15.

Gelfand, I.M. and M.A. Naimark [1943]. On the embedding of normed rings into the ring of operators in Hilbert space. *Mat. Sb.* **12 (54)**, 197-213.

Girard, J.-Y. [1972]. *Interprétation fonctionelle élimination des coupures dans l'arithmétique d'ordre supérieure.* Ph.D. thesis, Université Paris VII.

Gödel, K. [1930]. Die Vollständigkeit der Axiome des logischen Funktionenkalkuls. *Monatshefte für Mathematik und Physik* **37**, 349-360. Translated in van Heijenoort [1967], 582-591.

Gödel, K. [1938]. The consistency of the axiom of choice and of the generalized continuum-hypothesis. *Proceedings of the U.S. National Academy of Sciemces* **24,** 556-7.

Gödel, K. [1938a]. Consistency-proof for the generalized continuum-hypothesis. *Proceedings of the U.S. National Academy of Sciemces* **25,** 220-4.

Gödel, K. [1940]. *The Consistency of the Axiom of Choice and of the Generalized Continuum-Hypothesis with the Axioms of Set Theory.* Annals of Mathematics Studies No. 3. Princeton University Press.

Gödel, K. [1964]. Remarks before the Princeton Bicentennial Conference. In *The Undecidable,* Martin Davis, ed. Raven Press, pp. 84-88.

Goodman, N. and Myhill, J. [1978] Choice implies excluded middle. *Z. Math Logik Grundlag. Math* **24,** no. 5, 461.

Grayson, R.J. [1975] *A sheaf approach to models of set theory.* M.Sc. thesis, Oxford University.

Grayson, R. J. [1979] *Heyting-valued models for intuitionistic set theory.* In Fourman, M. P., Mulvey, C. J., and Scott, D. S. (eds.) *Applications of Sheaves. Proc. L.M.S. Durham Symposium* 1977. Springer Lecture Notes in Mathematics 753, pp. 402-414.

Halpern, , J.D. and Levy, A. [1971]. The Boolean prime ideal theorem does not imply the axiom of choice. *Axiomatic Set Theory.* Proceedings of Symposia in Pure Mathematics, Vol. XIII, Part I. American Mathematical Society, pp. 83-134.

Hamel, G. [1905]. Eine Basis aller Zahlen und die unstetigen Lösungen der Funktionalgleichung : $f(x + y) = f(x) + f(y)$. *Mathematische Annalen* **60**, 459-62.

Hartogs, F. [1915]. Über das Problem der Wohlordnung. *Mathematische Annalen* **76**, 436-443.

Hausdorff, F. [1909] Die Graduierung nach dem Endverlauf. *Königlich Sächsichsen Gesellschaft der Wissenschaften zu Leipzig, Math. - Phys. Klasse, Sitzungberichte* **61,** 297-334.

Hausdorff, F. [1914]. *Grundzüge der Mengenlehre.* (Leipzig: de Gruyter). Reprinted, New York: Chelsea, 1965.

Hausdorff, F. [1914a]. Bemerkung über den Inhalt von Punktmengen. *Mathematische Annalen* **75,** 428-433.

Henkin, L. [1949]. The completeness of the first-order functional calculus. *J. Symb. Logic* **14,** 159-166.

Henkin, L. [1954]. Metamathematical theorems equivalent to the prime ideal theorem for Boolean algebras. *Bull. Amer. Math. Soc.* **60,** 387-388.

Herrlich, H. [2002]. The axiom of choice hold iff maximal closed filters exist. *Math. Logic Quarterly* **49 (3),** 323-4.

Hessenberg, G. [1906]. *Grund begriffe der Mengenlehre.* Göttingen: Vandenhoeck & Rupprecht.

Hilbert D. [1926]. Über das Unendliche. *Mathematische Annalen* **95.** Translated in van Heijenoort, ed. From Frege to Gödel: A Source Book in Mathematical Logic 1879-1931, Harvard University Press, 1967, pp. 367-392.

Hodges, W. [1979]. Krull implies Zorn. J. London Math. Soc. **19,** 285-7.

Hoffman, K.H. and Lawson, J.D. *The spectral theory of continuous distributive lattices.* Trans. Amer. Math. Soc. **246**, 285-310.

Howard, P. [1975]. Łoś's theorem and the Boolean prime ideal theorem inply the axiom of choice. *Proc. Amer. Math. Soc.* **49**, 426-428.

Howard, P. and Rubin, J. E. [1998]. *Consequences of the Axiom of Choice.* American Mathematical Society Surveys and Monographs, Vol. 59.

Howard, W. A. [1980] The formulae-as-types notion of construction. In J. R. Hindley and J. P. Seldin (eds.), *To H. B. Curry: Essays on Combinatorial Logic. Lambda Calculus and Formalism,* pp. 479-490. New York and London: Academic Press.

Isbell, J. R. (1972) Atomless parts of spaces. *Math. Scand.* **31**, 5-32.

Jacobs, B. [1999] *Categorical Logic and Type Theory.* Amsterdam: Elsevier.

Jech, T. [1973]. *The Axiom of Choice.* North-Holland.

Jelonek, Z. [1993]. A simple proof of the existence of the algebraic closure of a field. *U Iagell. Acta Math. Fasic. XXX*, 131-132.

Johnstone, P. T. [1977] *Topos Theory.* London: Academic Press.

Johnstone, P. T. [1981] Tychonoff's theorem without the axiom of choice. *Fund. Math.* **113**, 21-35.

Johnstone, P. T. [1982] *Stone Spaces.* Cambridge University Press.

Johnstone, P. T. [1983] The point of pointless topology. *Bull. Amer. Math. Soc.(N.S.)* **8**, no.1, 41-53.

Johnstone, P. T. [2002] *Sketches of an Elephant: A Topos Theory Compendium, Vols. I and II.* Oxford Logic Guides Vols. 43 and 44, Oxford: Clarendon Press.

Kelley, J. L. [1950] The Tychonoff product theorem implies the axiom of choice. *Fundamenta Mathematicae* **37**, 75-76.

Kestelman, H. [1951]. Automorphisms in the field of complex numbers. *Proc. Lond. Math. Soc.* **(2) 53**, 1 - 12.

Kestelman, H. [1960]. *Modern Theories of Integration.* Dover.

Klimovsky, G. [1958] El teorema de Zorn y la existencia de filtros a ideales maximales en los reticulados distributivos. *Rev. Un. Math. Argentina* **18**, 160-64.

Kneebone, G. T. [1963]. *Mathematical Logic and the Foundations of Mathematics.* Van Nostrand.

Kunen, K. [1980]. *Set Theory.* North-Holland.

Kuratowski, K. [1922] Une méthode d'élimination des nombres transfinis des raissonements mathématiques. *Fundamenta Mathematicae* **3**, 76-108.

Lambek, J. and Scott, P. J. [1986] *Introduction to Higher-Order Categorical Logic.* Cambridge: Cambridge University Press.

Lang, S. [2002]. *Algebra.* Revised third edition. Springer.

Lawvere, F. W. [1972] Introduction to *Toposes, Algebraic Geometry and Logic.* Springer Lecture notes in Math. 274, pp. 1-12.

Lawvere, F. W. [1976] Variable quantities and variable structures in topoi. In A. Heller and M. Tierney, eds., *Algebra, Topology and Category Theory: a collection of papers in honor of Samuel Eilenberg.* New York: Academic Press, pp. 101-31.

Lawvere, F. W. and Rosebrugh, R [2003]. *Sets for Mathematics.* Cambridge University Press.

Leisenring, A.C. [1969]. *Mathematical Logic and Hilbert's ε-Symbol.* Gordon and Breach.

Lindenbaum, A., and Mostowski, A. [1938]. Über die Unabhängigkeit des Auswahlsaxioms und einiger seiner Folgerungen. *Comptes Rendus des Séances de la Société des Sciences et des Lettres de Varsovie* **31,** 27-32.

Löwenheim, L. [1915]. Über Mögglichkeiten im Relativkalkul. *Mat. Annalen* **76**, 447-470. Translated in van Heijenoort [1967], 228-251.

Mac Lane, S. and Moerdijk, I. [1994]. *Sheaves in Geometry and Logic: A First Introduction to Topos Theory.* Springer.

McLarty, C. [1988]. *Elementary Categories, Elementary Toposes.* Oxford University Press, 1988.

Maietti, M. E. [1999]. *About effective quotients in constructive type theory.* In Types for Proofs and Programs, International Workshop "Types 98", Altenkirch, T., et al., eds., Lecture Notes in Computer Science 1657, Springer-Verlag, pp. 164-178

Maietti, M. E. [2005]. Modular correspondence between dependent type theories and categories including pretopoi and topoi. *Math. Struct. Comp. Sci.* **15 6**, 1089-1145.

Maietti, M. E. and Valentini, S. [1999]. *Can you add power-set to Martin-Löf intuitionistic type theory?* Mathematical Logic Quarterly **45**, 521-532.

Malcev, A. [1941]. On a general method for obtaining local theorems in group theory. Translated in Malcev, A. *The*

Metamathematics of Algebraic Systems: Collected Papers, B.F. Wells, ed. North-Holland, 1971.

Mancosu, P. [1998]. *From Brouwer to Hilbert.* Oxford University Press.

Martin-Löf, P. [1975] An Intuitionistic theory of types; predicative part. In H. E. Rose and J. C. Shepherdson (eds.), *Logic Colloquium 73,* pp. 73-118. Amsterdam: North-Holland.

Martin-Löf, P. [1982] Constructive mathematics and computer programming. In L. C. Cohen, J. Los, H. Pfeiffer, and K.P. Podewski (eds.), *Logic, Methodology and Philosophy of Science VI,* pp. 153-179. Amsterdam: North-Holland.

Martin-Löf, P. [1984] *Intuitionistic Type Theory.* Naples: Bibliopolis.

Martin-Löf, P. [2006]. 100 years of Zermelo's axiom of choice: what was the problem with it? *The Computer Journal* **49 (3)**, pp. 345-350.

Moore, G. H. [1982]. *Zermelo's Axiom of Choice.* Springer-Verlag.

Myhill, J. and Scott, D.S. [1971]. Ordinal definability. *Axiomatic Set Theory.* Proceedings of Symposia in Pure Mathematics, Vol. XIII, Part I. American Mathematical Society, pp. 271-8.

Peremans, W., [1957] *Embedding of a distributive lattice into a Boolean algebra.* Indag. Math. **19**, 73-81.

Ramsey, F. P. [1926]. The Foundations of Mathematics. *Proc. Lond. Math. Soc.* **25,** 338-84.

Rubin, H. and Rubin, J. E. [1985]. *Equivalents of the Axiom of Choice II.* North-Holland.

Rubin, H. and Scott, D.S. [1954] Some topological theorems equivalent to the prime ideal theorem. *Bull. Amer. Math. Soc.* **60,** 389 (Abstract).

Russell, B. [1903]. *The Principles of Mathematics.* Cambridge University Press.

Russell, B. [1906] On some difficulties in the theory of transfinite numbers and order types. *Proc. London Math. Soc.* (2), **4,** 29-53.

Russell, B. [1908] Mathematical logic as based on the theory of types. *Am. J. Math.* **30**, 222-262. Reprinted as pp.150–82 in van Heijenoort [1967].

Russell, B. [1919]. *Introduction to Mathematical Philosophy.* Allen & Unwin.

Russell, B. and Whitehead, A. N. [1910-13]. *Principia Mathematica.* 3 vols., Cambeidge University Press.

Sambin, G. [1988] Intuitionistic formal spaces and their neighbourhood. *Logic Colloquium 88 (Padova, 1988),*261-265. Amsterdam: North-Holland.

Scott, D. S. [1954]. The theorem on maximal ideals in lattices and the axiom of choice, *Bull. Amer. Math. Soc.* **60,** 83.

Scott, D. S. [1966]. *More on the axiom of extensionality.* In Essays on the Foundations of Mathematics, Magnes Press, Jerusalem, 115-131.

Sikorski, R. [1948]. A theorem on extensions of homomorphisms. *Annales de la Societé Pol. de Mathématiques,* **21,** 332-35.

Skolem, T. [1920]. Logisch-kombinatorische Untersuchungen über die Erfüllbarkeit oder Beweisbarkeit mathematischer Sätze nebst einem Theoreme über dichte Mengen. *Videnskaps-selskapets Skrifter, I.,* 1 - 36. Translated in van Heijenoort [1967], 252-263.

Solovay, R. [1970] A model of set theory in which every set of reals is Lebesgue measurable. *Annals of Mathematics* **92,** 1-56.

Steinitz, E. [1910] Algebraische Theorie der Körper. *Journal für die Reine und angewandte Mathematik (Crelle)* **137,** 167-309.

Stone, M. H. [1936]. The theory of representations for Boolean algebras. *Trans. Amer. Math. Soc.* **40,** 37-111.

Stone, M. H. [1937]. Applications of the theory of Boolean rings to general topology. *Trans. Amer. Math. Soc.* **41,** 375-481.

Stone, M. H. [1940]. A general theory of spectra, I. *Proc. Nat. Acad. Sci. USA* **26,** 280-3.

Tait, W. W. [1994] The law of excluded middle and the axiom of choice. In *Mathematics and Mind,* A. George (ed.), pp. 45-70. New York: Oxford University Press.

Tarski, A. [1924]. Sur quelques théorèmes qui équivalent à l,axiome du choix. *Fund. Math.* **5,** 147-154.

Tarski, A. [1948]. Axiomatic and algebraic aspects of two theorems on sums of cardinals. *Fudamenta Mathematicae* **35,** 79-104.

Tarski, A. and R. Vaught [1957]. Arithmetical extensions of relational syatems. *Comp. Math.* **13,** 81-102.

Tychonov, A. [1935]. Über einen Funktionenräum. *Math. Annalen* **111,** 762-766.

Valentini, S. [1996]. A completeness theorem for formal topologies. *Logic and Algebra (Pontigniano, 1994),* 689-702. Lecture Notes in Pure and Applied Mathematics **180.** New York: Dekker.

Valentini, S. [2002]. *Extensionality versus constructivity.* Mathematical logic Quarterly **42** (2), pp. 179-187.

van Heijenoort, J., ed. [1967]. *From Frege to Gödel: A Source Book in Mathematical Logic, 1879-1931.* Harvard University Press.

Vitali, G. [1905]. *Sul problema della misura dei gruppi di punti di una retta.* Bologna: Tip. Gamberini e Parmeggiani.

Zermelo, E. [1904] Neuer Beweis, dass jede Menge Wohlordnung werden kann (Aus einem an Herrn Hilbert gerichteten Briefe) *Mathematische Annalen* **59** , pp. 514-16. Translated in van Heijenoort, *From Frege to Gödel: A Source Book in Mathematical Logic* 1879-1931, Harvard University Press, 1967, pp. 139-141.

Zermelo, E. [1908] Neuer Beweis für die Möglichkeit einer Wohlordnung, *Mathematische Annalen* **65** , pp. 107-128. Translated in van Heijenoort, *From Frege to Gödel: A Source Book in Mathematical Logic* 1879-1931, Harvard University Press, 1967, pp. 183-198.

Zorn, M. [1935] A remark on method in transfinite algebra. *Bull. Amer. Math. Soc.* **41,** 667-70.

Zorn, M. [1944]. Idempotency of infinite cardinals. *University of California Publications in Mathematics;* Seminar Reports (Los Angeles) **2,** 9- 12.

Index

www.ingramcontent.com/pod-product-compliance
Lightning Source LLC
Chambersburg PA
CBHW061149220326
41599CB00025B/4411